REGULATORS' REVENGE

REGULATORS' REVENGE

The Future of Telecommunications Deregulation

edited by Tom W. Bell and Solveig Singleton

Washington, D.C.

Library of Congress Cataloging-in-Publication Data

Regulators' revenge : the future of telecommunications deregulation / edited by Tom W. Bell and Solveig Singleton.
p. cm.
Includes bibliographical references and index.
ISBN 1-882577-68-X (cloth).—ISBN 1-882577-69-8 (pbk.)
1. Telecommunication—Law and legislation—United States.
2. Telecommunication—Deregulation—United States. I. Bell, Tom W.
II. Singleton, Solveig.
KF2765.Z9R44 1998
343.7309'94—dc21 98-36655
CIP

Printed in the United States of America.

CATO INSTITUTE
1000 Massachusetts Ave., N.W.
Washington, D.C. 20001

Contents

Preface

The latter half of the 20th century has witnessed extraordinary growth in the unregulated computer industry; by contrast, the heavily regulated telecommunications industry has lagged far behind. The Telecommunications Act of 1996 promised to deliver deregulation and free markets to the telephone, cable television, and broadcast industries. But did the act deliver? Critics assail it from all sides. Consumer groups complain that cable and telephone rates are rising instead of falling. Long-distance phone companies and cable television companies have not rushed to offer local telephone service, and local phone companies have not yet been freed to offer long-distance service. The Federal Communications Commission has grown larger than ever and continues to churn out volumes of regulation. Perhaps the act moved us a step closer to free markets in telecommunications—but we have not yet arrived.

The essays in this book describe innovative ideas for truly freeing markets in the wake of the act. The contributors draw on their expert knowledge about the aftermath of the 1996 act as well as on the progress of more radical telecommunications deregulation in Central and South America, New Zealand, Great Britain, Australia, and Canada. Their comments offer key insights into the role of antitrust law in telecommunications deregulation and the policy issues surrounding interconnection, universal service, spectrum auctions, and broadcast licensing. All the papers were first presented at the Cato Institute's Telecommunications Conference, "Beyond the Telecommunications Act of 1996," that was held in Washington on September 12, 1997. The substance of Alfred E. Kahn's paper was given as the luncheon address. The papers have been updated for 1998.

We thank all the contributors for their hard work in preparing these papers for publication. Some of the papers have been edited and updated substantially. Thanks to David Lampo and Elizabeth Kaplan of the Cato Institute for guiding the book through the publication process and to Jennifer DePalma for her research assistance.

Finally, we thank Ed Crane and David Boaz for their continuing support and shared wisdom.

Tom W. Bell and
Solveig Singleton

Introduction

Tom W. Bell and Solveig Singleton

Great Expectations

President Bill Clinton signed the Telecommunications Act of 1996 into law on February 8, 1996. The act is now over two years old. By most accounts, the 111-page act has proven something of a disappointment.

Consider some of the inspiring claims made about the act at the time of its birth, in early 1996: President Clinton, at the lavish signing ceremony for the act, said, "Today with the stroke of a pen our laws will catch up with our future. We will help to create an open marketplace where competition and innovation can move as quick as light."[1] Vice President Al Gore rhapsodized that the act charted "a new flight path to an entirely new world, a world in which we use technology to put us more directly in contact with each other, a world in which our ability to create and receive information will be limited not by the bounds of our technologies but only by the infinite boundaries of our imagination."[2] Such claims about the 1996 Telecom Act now sound a bit like Eisenhower-era predictions that atomic power would soon cost too little to meter and that we would commute via jetpacks. At the act's signing, Reed Hundt, chairman of the Federal Communications Commission, greeted it with a more revealing sort of enthusiasm: "The bill vests serious responsibilities in the FCC to make competition a reality in as many markets as possible. . . . The workload the bill will generate will require a significant commitment of Commission personnel and will stretch our limit."[3]

Not everyone held a rosy view of the 1996 act. An editorial in the *Wall Street Journal* opined that "the stuff of hopeful press releases "would not become reality for several years. The editorialists added that "the more immediate impact of the legislation will be more prosaic: It will spawn a lobbying frenzy at the doors of the FCC."[4] In retrospect, the *Wall Street Journal* seems to have gotten it right.

One commentator claims that consumer dissatisfaction with the act made it an "albatross" around the neck of former South Dakota senator Larry Pressler, effectively forcing him out of office.[5]

Perhaps competition, deregulation, and the expected benefits for consumers are coming, albeit more slowly than originally predicted. Telecommunications regulations have grown so incredibly complex over the last several decades that we can hardly expect overnight deregulation. What, then, can we reasonably demand of telecommunications reform?

First, going backward or sideways will certainly not do; markets should always move toward greater freedom. The elaborate universal service mechanisms established by the act contradict that goal. Some of the contributors to this book attribute a similar flaw to the act's interconnection provisions. Second, we should see some light at the end of the regulatory tunnel—however distant. Spectrum privatization would offer a critical first step. So would true deregulation of pricing at all levels—local residential service, access charges, everything. Third, serious regulatory reform must include long-term plans for closing down the FCC and returning most or all of its functions to the states or to the private sector.

By those standards, the Telecommunications Act of 1996 falls short of adequate deregulation. A review of funding and staffing trends at the FCC (Table 1.1) adds a quantifiable edge to this sharp assessment of the act.

Simply put, recent funding and staffing trends do not indicate that the FCC plans to close up shop and let the market blossom.

Or do they? Gigi Sohn of the Media Access Project criticizes the assumption that a growing FCC means that deregulation has stalled. The notion is that the FCC needs to do more now so that it can do less later. Analysts enjoy embracing paradoxes. Perhaps they will fasten their hopes to this one—the alternative is to admit that the act is failing and that the hard-fought battles that created it will have to be fought all over again.

We take a less forgiving view of the act, however. The FCC has not been trimming and cutting. Its regulations, not to mention its budget and staff, keep growing. No sunset of any sort looms for the FCC. Under both Commissioners Reed Hundt and William Kennard, the FCC has not only continued pursuing old missions but eagerly embraced new ones. Consider, for example, its recent ambitious

Table 1.1
FCC Funding and Staff, 1995–99

Fiscal Year	Funding	Personnel
1995	$184,232,000	2,130
1996	$185,709,000	2,060
1997	$189,079,000	2,032
1998	$186,514,000	2,100 (estimated)
1999	$212,977,000 (requested)	2,100 (requested)

Sources: For FY95: Office of Management and Budget, *Budget of the United States Government, Fiscal Year 1996—Appendix* (Washington: Government Printing Office, 1994), p. 975 (budget); *Budget of the United States Government, 1997—Appendix,* p. 978 (full-time equivalent personnel). For FY96: Federal Communications Commission, *62nd Annual Report, FY 1996* (FCC: Washington, 1997), p. 18 (funding); *Budget of the United States Government, Fiscal Year 1998—Appendix,* p. 1039 (full-time equivalent personnel). For FY97: *Budget of the United States Government, Fiscal Year 1998—Appendix,* p. 1038 (budget); *Budget of the United States Government, Fiscal Year 1999—Appendix,* p. 1074 (full-time equivalent personnel). For FY98: *Budget of the United States Government, Fiscal Year 1999—Appendix,* pp. 1073–74. For FY99: *Budget of the United States Government, Fiscal Year 1999—Appendix,* pp. 1073–74.

inquiries into children's educational television (resulting in costly new shows that, it turns out, children do not much care to watch), liquor advertising, and, now, free campaign advertising for politicians—a generous offering to the only people capable of threatening the regulatory status quo.

That leaves it up to us to revisit the Telecommunications Act of 1996, to critique it, perhaps even to praise it, but at any rate to fix it. Because proponents of the act described it as "deregulation," both apologists for and critics of the act now mischaracterize its salient flaws as market failures rather than government ones. For example, rising cable rates—a predictable response to loosening price caps that starved the industry of capital—have raised calls for new regulation. Similar misapprehensions lie behind the outcry over recent mergers in the telecommunications industry—mergers that simply reflect the need for capital to compete in global markets and the drive for economies of scale and scope. By deregulating impartially and imperfectly, the act threatens to invite a revenge of the regulators.

The Transition to True Deregulation

The papers in Part I describe alternative general frameworks for free telecommunications markets. The authors focus on pitfalls that regulators should avoid, assess the role of antitrust law, and examine how radical deregulation has proceeded around the world.

In chapter 2, Alfred E. Kahn, the Robert Julius Thorne Professor of Political Economy, Emeritus, Cornell University, describes the circumstances that allowed regulators to deregulate the airline industry quickly and with minimal interference. The airlines had indulged in a binge of acquiring new capacity, just as demand became sluggish because of the 1973–74 recession, so all that was needed was for regulators to get out of the way and watch prices fall. So the first lesson of deregulation, for Kahn, is to be lucky. The second lesson, drawn from his experience with airline and trucking deregulation, is that opening the markets to new entry made little sense unless pricing was also deregulated, and micromanaging the process creates more distortions than it cures.

Telecommunications, Kahn argues, will require somewhat more regulatory intervention because incumbents in local telephone markets control facilities essential to their competitors' success and because rate regulation and universal service obligations continue. So he believes that regulators will have an essential role in telecommunications. On the other hand, the role tempts them to indulge in political opportunism by making rules that will yield a quick crop of live "competitors" regardless of whether those rules handicap competition and heavily subsidize new entrants. The FCC, he concludes, has fallen into the trap in pricing both resale and unbundled network elements.

For Kahn, prices ought to be set as they would be in a market, that is, on the basis of the actual costs of the incumbent companies. That encourages and rewards new entrants who can provide services by building their own lower cost facilities. The FCC's heavily discounted pricing regime eliminates both the incentive and the reward and simply allows the new entrant to buy facilities from the incumbent.

In chapter 3, Peter W. Huber argues for a comprehensive reassessment of the justifications for federal regulation of telecommunications. Huber's powerful statement makes clear that he regards the regulations, enforcement efforts, and very existence of the FCC as constitutionally suspect. Given such a dubious pedigree, it comes as little surprise that the FCC has issued swarms of meddlesome rules that interfere with private conduct and freedom of speech.

Huber calls for abolishing the FCC and relying on common law processes to regulate the rapidly growing world of telecommunications. Huber would have courts apply not only the familiar tenets of contract, tort, and property law to telecommunications disputes, however. He also would have courts interpret and apply the legislation that created antitrust and copyright laws. In either case, it matters most that courts build the law from the bottom up, from general principles to specific rules. Only that sort of flexible and decentralized process, Huber argues, will allow the evolution of the law to match that of telecommunications.

Pablo T. Spiller's and Carlo G. Cardilli's view on the proper role of general antitrust principles in moving toward free telecommunications markets, expressed in chapter 4, contrasts markedly with Huber's. Spiller and Cardilli describe the radical telecommunications reforms that have taken place in Chile, New Zealand, Australia, and Guatemala. They conclude that leaving interconnection disputes to be resolved in the courts under general laws results in delay and burdensome regulation. They favor instead the type of regime chosen by Guatemala, under which interconnection regulation rules are established by a telecommunications regulatory agency that has much less discretion than the FCC has in the United States. Disputes over the terms of interconnection are to be resolved under final-offer arbitration, to discourage companies from using lengthy appeals to regulators and to the courts to beat out the competition, as is often done in the United States.

Spiller and Cardilli draw other lessons from their extensive knowledge of the deregulation process in other countries. In particular, they note that, even with the delays caused by litigation in Chile and New Zealand, facilities-based competition in the local exchange has increased enormously, especially in Chile. That, they note, shows that the local exchange is not, as skeptics claim, a natural monopoly.

They also make clear that other countries have been well aware of the dangers of giving new entrants into the local exchange too generous rights of resale and interconnection. Guatemala, in particular, will try to encourage rapid new entry into telecommunications markets by allowing new entrants to sell unbundled elements of the incumbent's networks—but the unbundling rights will last a limited number of years. The idea is to have the best of both worlds—rapid new entry, but not at the expense of facilities-based competition.

Interconnection and Infrastructure in a Free Market

The papers in Part II assess the interconnection rules established under the Telecommunications Act of 1996—not the acrobatics of pricing calculus performed by the FCC but the general principles underlying the FCC's deliberations.

Tom Tauke, senior vice president for government relations with Bell Atlantic, provides in chapter 5 an overview of the way to proceed toward competition without changes to the Telecommunications Act of 1996. Tauke attributes the perception that the act is a failure partly to impatience and unrealistic expectations. But he also points to the need for rate rebalancing, particularly in the local exchange, as new entrants will shun markets where prices are held low. Tauke also argues that Congress intended interconnection relationships between incumbent phone companies and new entrants to be negotiations and notes that many such negotiations have been concluded successfully. Sometimes, though, negotiations have deteriorated into regulatory posturing.

Tauke concludes by describing four regulatory principles to help move the regulatory regime under the Telecommunications Act of 1996 toward free markets. The third of those is of particular interest; he recommends that regulators keep innovative new services free from the old regulatory apparatus.[6]

Henry Geller of the Markle Foundation expresses in chapter 6 his view that telecommunications reforms in the United Kingdom and Canada "call into question the course of interconnection regulation taken by the United States and militate strongly for a modest course correction at this time."

Both Canada and the United Kingdom have accepted that resale and interconnection are important to new entry into local phone markets. But each country also has recognized that too generous interconnection and resale rights can discourage facilities-based competition. Thus, the United Kingdom has not given new entrants unbundling rights. Canada will mandate unbundling of local loops only for a limited time.

Geller encourages regulators in the United States to recognize that too expansive interconnection policies can deter the build-out of new infrastructure. He views section 706 of the act, which allows the FCC to use forbearance to encourage the deployment of advanced networks, as an opportunity for the FCC to refrain from drawing advanced and innovative services into the interconnection regime.

In chapter 7, Solveig Singleton, director of information studies at the Cato Institute, asks about interconnection in the long run. The current consensus that mandatory interconnection is necessary to promote new entry into the local exchange hides fundamental disagreements about the nature of competition in telephone services. Will it ever be possible to do away with mandatory interconnection entirely?

She next argues that the substantial drawbacks of mandatory interconnection policies suggest an answer to that question. Those drawbacks include the inescapable problems of pricing interconnection "correctly," and the ill will the regime fosters between different sectors of the telephone industry. Because interconnection now and in the future promises thickets of thorny regulation and little free-market fruit, we should begin to consider now how mandatory interconnection rules should be eliminated. She concludes with some suggestions for scaling back the rules.

Peter K. Pitsch of Pitsch Communications argues in chapter 8 that political constraints mean that the only way to deregulate telecommunications is for the public and policymakers actually to see new entry into telephone markets. He supports this by noting the history of deregulation in long-distance telephone markets. And he concludes that the surest road to new entry is unbundling and rate rebalancing.

Pitsch further argues that the prices of the unbundled network elements should be based on forward-looking cost estimates. The estimated costs should be those of the "incumbent local telephone company's network, assuming current wire center locations," rather than the costs of a hypothetical network. He advocates allowing carriers to recover common costs (the overhead costs the incumbent accrues by offering unbundled elements) by charging each subscriber using the incumbent network (either the incumbent's customer or the reseller's) an end-user charge. To minimize regulatory interference, he urges that the prices for unbundled network elements be placed under price caps, allowing negotiations to proceed under those caps.

Speculating in Spectrum: The Future of Property Rights

In chapter 9, the first chapter of Part III, Thomas W. Hazlett describes how the Telecommunications Act of 1996 largely sidestepped the hard work of deregulating the wireless sector. Although economists roundly deride the FCC's present methods of allocating rights to the electromagnetic spectrum as inefficient and anticompetitive, reform will not come easily. Hazlett attributes the problem to the very structure of the regulatory system, which legally obligates agency personnel to allocate spectrum in the "public interest," via administrative procedures.

Only fundamental reform will save wireless telecommunications from a regulatory chokehold. Hazlett calls for opening the spectrum to free access and use, subject only to a bar on interference with existing transmissions. He thus criticizes as too restrictive the current practice of auctioning off mere licenses to use certain portions of the spectrum in certain ways. Such auctions not only give the FCC too much control over spectrum allocation; they also encourage lawmakers to artificially restrict spectrum access to inflate receipts. Hazlett recounts how politics has likewise corrupted regulation of the broadcast industry. Real deregulation of wireless telecommunications, he concludes, must cut Washington out of the loop and treat spectrum like any other resource developed, sold, and used in the free market.

In chapter 10, Stanley S. Hubbard gives a lively firsthand account of his family's struggle to develop the broadcast spectrum. Their pioneering efforts amply demonstrate that spectrum has no value in the abstract. Broadcast television, for instance, at first met widespread skepticism and indifference. It took entrepreneurs like the Hubbard family to risk their fortunes, devote their labors, and make broadcast television what it is today: a hugely successful industry and a staple of American culture. A similar combination of public doubt and private vision has driven Hubbard's development of the direct broadcast satellite spectrum.

Hubbard argues from practical experience that settlers on the frontier of the electromagnetic spectrum deserve property rights no less than settlers on America's terrestrial frontiers did. In each case, assuming the risks and burdens of developing an unused territory should suffice to win good title to it. Spectrum developers have an even stronger case for homesteading rights, in fact, since they neither displace natives nor upset ecosystems. Rather, they turn something that has no intrinsic value into services citizens can hardly imagine living without.

In chapter 11, Evan R. Kwerel and John R. Williams offer a comprehensive overview of how to implement a market-based spectrum policy. They call for the FCC to license spectrum to specific parties in perpetuity and then get out of the way, leaving the licensees free to use spectrum in any noninterfering manner or reassign it, in whole or part, to third parties. Kwerel and Williams counsel the government not to reserve spectrum, reasoning that private parties will buy up and set aside spectrum if doing so makes economic sense. The authors would moreover apply their market-based reforms to nearly every bit of spectrum, across all frequencies and geographic locations.

Kwerel and Williams describe several steps that the FCC has already taken toward a market-based spectrum policy (as well as a few missteps away from it). Do market failures stand in the way of further reforms? The authors examine the question and generally find the risks of market failure inconsequential, less worrisome than the risks of government failure, or subject to relatively easy remedies.

Eli M. Noam argues in chapter 12 for a paradigm shift in how we think about deregulating spectrum allocation. Whereas the current policy debate pits a regulatory assignment model against an auction model, Noam argues for an "open access" model in which those who need spectrum buy access to it piecemeal, rather than buying exclusive licenses to entire spectrum bands. He likens this system to a toll road that sells access on a per vehicle basis and controls congestion by moderating prices.

Noam defends open access as more constitutionally sound than systems that exclusively license access to the wireless media. While he recognizes auctions as an improvement over regulatory assignment of spectrum, Noam argues that auctions inevitably put revenue receipts before considerations of economic efficiency and freedom of speech. He describes some historical precedents for the open access alternative, outlines the main features of a modern implementation, and advocates that we begin experimenting with that promising new approach to deregulating the spectrum.

Information Have-Nots and Industrial Policy

Part IV begins with Gigi B. Sohn, director of the Media Access Project, arguing in chapter 13 that government involvement in the form of universal service subsidies can make markets work better and therefore help free markets to succeed. Sohn claims that the main item of expense for schools, libraries, and other groups that need more access to telecommunications is not hardware but service.

Sohn argues that government involvement is necessary to support access to telecommunications services because of the cost structure of telecommunications markets. The market will provide service only where there is profit to be made; the costs structure of telecommunications markets makes it unlikely that markets will provide universal service without government help. She adds, "I agree strongly with libertarians that subsidies must not be hidden. If government is going to give out money, it should be subject to public disclosure, not buried." She concludes, however, that universal service subsidies do make us all better off, even though there is some misallocation.

Lawrence Gasman, a senior fellow at the Cato Institute and president of Communications Industry Researchers, in chapter 14 strongly disputes the justifications offered for universal service subsidies. He describes the interest-group politics behind the spread of support for universal service, and he shows how the FCC failed to check that political dynamic by declining to place any limiting principles on the growth of universal service.

Gasman then refutes traditional justifications for universal service. First, he notes the high penetration figures for advanced telecommunications technologies in low-income households, suggesting that even low-income families can afford telecommunications services by careful budgeting. Gasman then attacks the "network externality" justification for universal service and the peculiar idea that urban and rural users should get the same phone service for the same price while in other respects urban and rural areas differ markedly.

Gasman offers an alternative to government universal service subsidies: freeing markets to spur innovation and competition. If entrepreneurs are freed to pursue profits, they will innovate to find new ways of lowering costs and serving underserved markets, for example, those in rural areas.

In our final chapter, 15, Bill Frezza of Adams Capital Management describes the conflict between innovation and universal service. The 1996 act's universal service regime creates a massive network of more or less hidden taxes, inefficient pricing, and complicated distribution mechanisms. Yet policymakers agree that the Internet and other advanced technologies should remain unregulated. Frezza predicts that the natural result of that will be a flow of capital and innovation to unregulated technologies.

Thus, for Frezza, innovation will always conflict with any attempts to establish a fixed regulatory structure for universal service. But the spirit of the process will remain the same: "an armed and surly mob of rural homeowners, underpaid librarians, and teachers' union shop stewards prepared to knock you over the head and yank a buck out of your wallet every time they want to make a phone call."

Conclusion

What lessons can we draw from this collection of papers? The contrasting views of Huber and Spiller present a fork in the road to telecommunications deregulation. We could proceed toward free markets under general antitrust rules and the common law, leaving courts to resolve interconnection and other complicated issues. Huber suggests that sort of path, which New Zealand and Chile have actually taken. Or we could follow Spiller's model of allowing competition to grow under a regulatory agency that has much less discretion that the FCC, the sort of path that Guatemala chose. Even Spiller's data suggest, however, that despite their reliance on relatively slow judicial processes, New Zealand and Chile eventually developed vigorously competitive telecommunications markets.

Both Huber's and Spiller's models thus deserve serious consideration. Each model requires far less regulation than the FCC currently administers. Each likewise counts on a legislature more committed to free markets and radical reform than Congress can boast of being. Lacking those prerequisites to real deregulation, the United States continues to wallow in outmoded regulatory models. Meanwhile, more innovative countries blaze new paths to free telecommunications markets.

From the interconnection panel, we garner more clues about how to deregulate telecommunications. Interconnection remains a stubborn problem—an assessment that Kahn underscores with illustrations from airline and trucking deregulation. Pitsch strongly argues that deregulation will follow only after new entry and rate rebalancing. But does luring new entrants require interconnection terms as favorable as those offered under the FCC's current scheme? Examples from Britain and Canada suggest not; facilities-based entry still merits encouragement. In the long run (say, 10 years), however, we must decide whether we want mandatory interconnection indefinitely. Answering yes will cast us into a losing battle of price regulation, manipulation of the regulatory process, extraordinary ill will between competitors, and—worst of all—the prospect of entire networks and systems of switches designed in response to regulation, either to support it or to subvert it. We should thus say no to long-term mandatory interconnection.

The spectrum panel thoroughly discredits the notion that the FCC should parcel out tightly limited rights to the spectrum on the basis of vague public interest standards. That consensus should come as no surprise, since most economists and many policy analysts have come to hold similar views. Somewhat remarkably, however, the panelists also roundly criticize spectrum auctions—the reigning fad in telecommunications deregulation—as too timid. Hazlett and Kwerel and Williams call for the FCC to put spectrum on the market and then step out of the way. Hubbard asks why those who develop unused spectrum should not win good title to it, while Noam asks whether we could do without licensing exclusive rights to whole bands of spectrum. Taken together, these approaches to deregulating spectrum allocation would give more play to market forces; show greater respect for common law property, tort, and contract principles; and decrease political control of the wireless media.

The universal service panel presents a conflict between Sohn's vision of what subsidies aim to accomplish and the political and market realities described by Gasman and Frezza. Gasman's arguments cut to the heart of the justifications for universal service. Subsidies have garnered support for political, not logical or economic, reasons. The universal service emperor wears no clothing. Its primary economic justification (the network externality) falls apart on close scrutiny. Because telecommunications markets do not have immutable cost structures, the best answer to the problems of high-cost areas and services is to let innovation proceed. Competition is the consumer's best friend, and price averaging and subsidies are the enemies of effective competition.

Notes

1. White House, Office of the Press Secretary, "Remarks by President Bill Clinton and Vice President Al Gore at Signing of Telecommunications Reform Act of 1996, February 9, 1996."
2. Ibid.
3. "Statement by FCC Chairman Reed E. Hundt Regarding Passage of the Telecommunications Act of 1996, February 8, 1996," available at 〈http://www.fcc.gov/Speeches/Hundt/spreh603.txt〉.
4. "Mr. Bell's Legacy," editorial, *Wall Street Journal*, February 7, 1996, p. A14.
5. Kirk Victor, "Call Waiting," *National Journal*, January 31, 1998, p. 234.
6. See Lawrence Gasman and Solveig Bernstein, "A 'Firewall' to Protect Telecom," *Wall Street Journal*, March 27, 1997.

Part I

The Transition to True Deregulation

1. Resisting the Temptation to Micromanage: Lessons from Airlines and Trucking

Alfred E. Kahn

I would like to begin with two disclosures: First, my life did not begin—or end!—with the deregulation of the airlines. Nor, second, do I believe that all wisdom about the proper role of government begins and ends with that experience. A little regulation could have saved the 250 lives that were lost because of the overloading of a ferryboat in Haiti on September 8, 1997.

The historical fact is, however, that while there were significant deregulations before that of the airlines—for example, of stock exchange brokerage commissions in 1976—the airline experience played the critical role in a development of truly historic proportions.

That occurred partly because it was the beneficiary of a serendipitous combination of circumstances: the industry in the preceding decade had engaged in a binge of acquiring jets and jumbo jets, greatly expanding its capacity just as the growth in demand decelerated, partly because of the 1973–74 recession. Its load factors averaged only around 53 percent in the 13 years before deregulation, and returns on equity averaged only 3 to 4 percent in the early 1970s. With all those lovely, empty, zero marginal cost seats, all it really took to produce huge benefits to the flying public was for us to get out of the way and let the companies follow their natural inclinations, under pressure of competition, to find ways of filling them.

As a result, load factors of the trunk airlines increased steadily, reaching 69 percent last year. With the recovery of the mid-1970s, we experienced the happy combination of both sharp reductions in real average yields and sharply increased profits for the companies.

Alfred E. Kahn is the Robert Julius Thorne Professor of Political Economy, Emeritus, at Cornell University and special consultant with National Economic Research Associates, Inc.

Had deregulation occurred instead in the circumstances of the early 1990s, when, after another binge of capacity acquisitions, the industry lost more money than it had made over its entire previous history, both Congress and the public would have been less willing to emulate that experience by deregulating other industries. Instead, we enjoyed the happy experience of the incipient deregulation proving good, in obvious ways, for the public and for the airlines as well. I am convinced that we never would have succeeded in deregulating trucking and the railroads two years later—one of our few achievements in my unfortunate tenure as Inflation Czar—had it not been for the happy way in which the airline experience worked out.

If the first lesson I would draw from the airline experience is "be lucky; come in at what turns out to be the right time and the right place," there are others, more fundamental, I think, that are turning out to be poignantly relevant to the case of telecommunications, and it is those lessons that I propose to draw today.

I have told the story many times of how we tried, for what seemed to be good reasons, to move gradually in deregulating the airlines, in effect micromanaging the process, partly out of sheer intellectual caution but also because we were eager not to be regarded as ideologues and were sensitive to the risks of frightening both Congress and the financial markets.

For these reasons, we started to go through the more than 600 applications for new route authority that had accumulated during the preceding half decade in the traditional way, case by case. Then, as we proceeded to grant individual applications of carriers to enter individual cities or routes, we began to encounter protests by incumbent carriers against our letting airlines in to compete with them while not giving them an equivalent opportunity to take advantage of opportunities elsewhere.

What we failed to realize, at first, was that individual routes are not separate entities but components of complex networks. What the carriers required—all of them—was total freedom to reconfigure their entire route structures. Proceeding route by route and case by case clearly was not going to work. That recognition was confirmed when our chief administrative law judge informed us, in our first major proceeding, the Midway case, that it would take 26 months to go through the applications in that one proceeding in the traditional way—city by city, applying the traditional public convenience and necessity criteria. Here, we realized, was a recipe for total frustration.

Second, as believers in competition, we decided from the outset to make the new entries that we were authorizing not mandatory but permissive. It soon became clear that we were, once again, guilty of partial liberalization: the incumbent carriers protested our doing so while still requiring them to remain in those markets. Though the obligation to serve was minimal in the number of flights per week required to retain a license, it became clear that freedom of entry without freedom of exit is not the same thing as free competition.

Third, even more sobering, we began to encounter the reality that routes with only one carrier had, on average, load factors—the average number of available seats sold to fare-paying passengers—in the range of 70 percent; that as one moved to routes with two carriers, that average dropped about 10 points, and with three carriers, another several points. It looked as though we were substituting inefficient oligopoly for efficient monopoly.

It did not take us long to realize that the problem was that we were deregulating entry but not prices. At their previously artificially sustained levels, carriers could break even by selling on average only some 50 percent of their seats. On the longest hauls, for fares that were deliberately held far above cost in an attempt to subsidize shorter routes, our economists assured us, carriers could break even filling only one-third of their seats. Forbidden the ability to compete in price, carriers competed by scheduling more and more flights on the long-distance routes. With price competition forbidden to drive break-even load factors up, the carriers resorted to scheduling competition, pushing achieved load factors down to those ridiculously low break-even levels. So it became clear to us, eventually, that freedom of entry had to be accompanied by freedom to compete in price instead of, wastefully, through density of scheduling or other such nonprice means.

In short, we came to the realization before very long that micromanaging the process creates more distortions than it cures. In that realization, we began to have the assistance of some of the major airlines, who had theretofore been the principal defenders of regulation. When they began to say to us, in effect, "If you are not going to protect us from competition, get out of our way," the process had taken on its own, sufficient momentum.

We had a similar experience just a few years later with deregulation of trucking. In deregulating entry Congress was unwilling to eliminate the obligation of carriers to file tariffs and adhere to them, largely for fear that in the absence of that obligation shippers would simply be unable to make intelligent choices. Just as in the case of the airlines, however, travel agents and computers were available to help travelers find their way through the maze of fares, so in the case of freight, the number of traffic brokers in the United States grew rapidly from the hundreds to the thousands, to fill that need.

The tariff-filing requirement was widely ignored, once carriers were freed to set their own rates and to negotiate long-term contracts with individual shippers. Retention of the tariff-filing requirement had embarrassing consequences many years later, as large numbers of truckers went into receivership under pressure of competition. Lawyers acting on behalf of the truckers' creditors sought out the favored shippers and presented them with bills for illegally received rebates totaling $32 billion, according to the estimates of the Interstate Commerce Commission (ICC). It took an additional intervention by Congress, in the form of the Negotiated Rates Act, to exempt shippers from those obligations unless and until the ICC found the officially filed rates reasonable. The new law brought to a close what threatened to be an extremely expensive lesson in the dangers of only partial economic deregulation.

Turning to Telecommunications Deregulation

I turn now to the case of telecommunications, observing incidentally that much of what I will have to say applies equally to the case of electric power. The situation in both cases is much more complicated than the airline one, where we discovered that the best thing was simply to get out of the way. In more traditional public utility industries, in contrast, there remains the necessity of protecting captive ratepayers until competition becomes sufficiently effective. The feasibility of competition is itself dependent upon rivals of the public utility companies having access, on nondiscriminatory terms, to essential facilities that are under the control of the incumbent companies—electric transmission and local distribution facilities, for example, and the local networks of the incumbent telephone companies, all of which continue, at least for the time being, to have the characteristics of natural monopolies. Moreover, the incumbent utility companies themselves continue to have unique obligations. Those obligations include extending service ubiquitously at rates that continue to be regulated and, especially in the telephone case, incorporating a gross regulatorily dictated cross-subsidization of basic exchange rates.

In these circumstances, simple deregulation is not possible:

- local monopoly continues to be ubiquitous;
- the still-regulated utility companies need reasonable opportunity to recover the costs of their continuing public utility obligations and regulatorily imposed cross-subsidizations; and
- would-be competitors require protections similar to those provided by the antitrust laws.

Just as efficient competition must be conducted on the basis of the relative efficiencies of the several contenders, so also must it be unbiased by the exercise of continued public utility monopoly power to exclude rivals from the opportunity to compete on that same basis.

Immensely complicating the problem of deregulating the telephone industry are the distorted rate structures upon which regulators continue to insist. For obvious reasons, when this market is opened to competition, it concentrates on the services whose rates have been deliberately overpriced to generate the cross-subsidies—estimated in the range of $20 billion to $25 billion a year—necessary to recover the costs of the other obligations. It does not take an economist to certify which are the rates that have been inflated in this way: all one need do is look at the markets on which competitors have concentrated. They are, first, long distance, both interstate and intrastate; then, access to the local telephone companies by the long-distance companies, at levels authorized by the Federal Communications Commission and the state commissions, after AT&T was broken up, sufficiently inflated above cost to recover from long-distance carriers the multi-billion-dollar contributions previously obtained directly from long-distance customers.

In those circumstances, it should not be surprising to learn, the second market that competitors have entered has been the provision of access to the local exchanges: every major metropolitan area in the country now has a competitive access provider, offering direct connection between long-distance companies and ultimate, particularly business, customers.

The third market into which competitors have flocked has been the provision of local dial-tone service to business customers, particularly in concentrated metropolitan areas, where charges are also grossly inflated to hold down basic residential rates.

In those circumstances, the central task confronting regulators has been to figure out how simultaneously to permit the incumbent companies recovery of their legitimately incurred costs—typically by charging rates far above costs for access to their essential facilities—and permit the entry of efficient competitors and discourage the entry of inefficient ones. The task has brought to the fore, over the last decade or so, the need for rules of competitive parity or efficient component pricing, in the formation of which I have been a significant participant—all of them necessitating continued regulation in the telephone and electric cases that was not required in the cases of the airlines and trucking.

Unfortunately but predictably, that essential, continuing role for regulators has provided the continuing occasion and justification for them to succumb to their high inherent marginal propensity to micromanage, as well as to practice political opportunism. Regulators thus have retained the inexcusably distorted, politically popular, but economically catastrophic across-the-board subsidization of basic residential rates and handicapped the competitive process to be able to show quick results in the form of live competitors, regardless of whether they are entitled to survive on the basis of their efficiency alone.

Before spelling out that indictment, I want to single out as an exception Mark Fowler, who, as chairman of the FCC in the 1970s, both recognized the distortion of the rate structures over which he presided (in carrier access charges inflated to continue the flow of subsidy from interstate to local subscription) and the distortions of access competition that it encouraged, and proposed a full rebalancing of access charges, on the one side, and basic residential rates—in subscriber line charges—on the other. He was forced by political pressure to settle for half a loaf; but even that was a profile in courage, for which I salute him. I confess that in so doing I am modestly saluting myself, because as chairman, in the mid-1970s, I persuaded the New York Public Utility Commission to take the first such modest step, raising basic residential charges (while retaining a bare "basic budget" option) and reducing intrastate toll.

Before proceeding to describe the mess created by that attempt to micromanage the introduction of competition while retaining gross politically dictated cross-subsidization, I am constrained to make two disclosures.

The first is that the intensity of my irritation with recent policies of the FCC is explained in some measure by the fact that they ignored my advice. Worse, they ignored a long footnote in my book in which, drawing upon earlier writing by an eminent economist, William Fellner, I warned them against the blank-slate version of total element long-run incremental cost (TELRIC) that they subsequently endorsed as the basis for the prescribed pricing of unbundled network elements.

The second disclosure is that while I was for six years a member of AT&T's first National Economic Advisory Council, I have in recent years been representing the Regional Bell Operating Companies in various proceedings at both the state and federal level, dealing with the same issues.

A Brief Characterization of Regulators' Errors

The limits of time permit me only briefly to characterize the major ways in which I feel regulators generally and the FCC in particular have fallen into the two traps of political expediency and economic error. The first is that they have engaged in inexcusably gross handicapping of the competitive process, committing the familiar error of restricting competition in the interest of protecting competitors, regardless of their relative efficiency. Specifically, encouraged by the Telecommunications Reform Act, they have extended to competitors or would-be competitors of the local exchange carriers (LECs) preferences, protections, and entitlements that go far beyond what antitrust principles would dictate. That is to say, under their inescapable mandate to ensure fair competition, they have taken substantial steps in the direction of illegitimate protections and preferences for competitors, regardless of their relative efficiency.

The Telecommunications Act itself is in important measure responsible for this outcome. But the FCC itself has interpreted the act as requiring the incumbent telephone companies to make available to their competitors, not just services that would qualify under the antitrust laws as genuinely monopolistic, essential facilities, but all network elements for which it is technically feasible to provide unbundled access. I know of no antitrust case or judgment, even against the most egregious violators of section 2 of the Sherman Act, that has gone so far. And, closely related, the FCC and some regulatory commissions—recognizing that full-fledged entry, offering all the services customers might wish to buy bundled, is likely to be infeasible—have required incumbent companies to make available to competitors all or any combination of their retail services at prescribed discounts in the 17 to 25 percent range. This is far in excess of their actual avoided costs and therefore inefficiently protective of competitors and promotive of competition.

Observe the subtle way in which this prescription by the FCC reinterprets the provision of the act, which requires only that the local companies make those retail services available at a discount reflecting their own avoided costs. The FCC instead prescribes discounts equal to its estimate of the average total cost of performing the entire retail function, on the ground that that would be the cost incurred by a new reseller. Following the FCC's own logic, the incumbent LECs would actually avoid costs equivalent to the prescribed 17 to 25 percent of their retail prices only if they got out of the retailing business entirely and became pure wholesalers: those are the incremental costs that purport to be measured by TSLRIC. But no one seriously maintains that those are the costs that the local telephone companies will actually avoid when, while *remaining* in the retail business, they make *some portion* of their output available to resellers at the prescribed discounts. So we have here a guaranteed margin for resellers much wider than the costs they actually save the incumbent companies by taking over a portion of their retail business.

Which of the two is the proper discount if efficient competition is to prevail? Not surprisingly, the answer was supplied by the late Nobel laureate William Vickery: it is the cost of the particular increment (or decrement) of sales involved in the transaction actually contemplated. Nobody is suggesting that the local telephone companies are going to get out of the retail business entirely, or that the resellers are going to take it over entirely. It is not surprising, therefore, that when the former come in with estimates of their avoided costs, the estimates fall to 5 percent or less. So basing the discounts on TSLRIC, rather than the relevant long-run incremental costs (LRIC), offers an enormous subsidy to new entrants and distorts the competition between them and the incumbents.

But, one may say, a new entrant into retailing in competition with an incumbent LEC will surely incur TSLRIC. How then can competitors enter at all if their discounts are only at the much lower LRIC level? The simple answer is that the most likely potential entrants, the companies already demanding the right to buy the retail services of the incumbent companies at prescribed discounts, are companies like AT&T, MCI, MFS, Brooke, TCG, and cellular service providers, all of whom are already retailing communications services—the relevant cost to whom is the cost of adding some services to their existing retail mix—in other words, that same LRIC rather than TSLRIC. They do not need the subsidy of a big margin, equivalent to the total cost of retailing as compared with not retailing. They need a margin equivalent only to the incremental cost to them of retailing the additional services: which is, if they are just as efficient as the incumbents, the discount that would represent the actual avoided costs of the latter companies as well.

So in its zeal to encourage competition, the FCC has prescribed discounts far greater than necessary to permit the entry and survival of equally efficient competitors—that is, competitors with marginal costs no higher than those of the incumbent telephone companies.

Moreover, the FCC and state commissions have interpreted the resale obligation without apparent limit of time or scope. Their conception of competition—that individual competitors must turn over to their rivals, at prescribed full-service incremental cost discounts, every product or service that they develop, past, present, and future—would surely make the eminent economist Joseph Schumpeter turn over in his grave.

I turn briefly to my other major quarrel with the FCC, and that is its prescription of something approximating the abominable TELRIC as the basis for the charges for unbundled network elements.

One assumption underlying that prescription is unexceptionable—namely, that efficient prices should be based on forward-looking costs. As I wrote two volumes on the efficiency of marginal-cost pricing, I can hardly argue against that proposition. But I never dreamed, when I asserted that unexceptionable principle, that some regulatory commission would assert that efficient prices should be based, not on the actual marginal or forward-looking costs of real-life suppliers, but on what it determines, in its own wisdom, to be the costs of some hypothetical most efficiently designed network, constructed from scratch, by an entrant writing, as it were, on a blank slate—TELRIC—BS.

What Professor Fellner pointed out 30 years ago was that the assumption behind the prescription—namely, that that is the level at which competition would set prices (at long-run marginal costs, with "long run" defined as the theoretical extreme, in which all costs are variable) is incorrect. No one, he pointed out, would, in the context of continuing technological progress, invest in constructing a new facility embodying the most recent version of a dynamic technology, the moment the progressively declining real cost fell below the market price. That facility would be obsolescent tomorrow and obsolete the day after. For that reason, as he put it, investors would practice what he called "anticipatory retardation": they would invest in new facilities only after the costs of doing so had fallen sufficiently *below* the market price to enable them to earn high returns in the early periods to compensate for the progressive obsolescence of the facility thereafter.

Jerry Hausman has translated that requirement into an estimate that, to compensate for the risk of obsolescence, investors would have to be promised the ability to recover their investments at high depreciation rates and to obtain rates of return in the early period two to three times the traditional regulatorily determined costs of capital.

Observe the effect of prescribing rates based on the hypothetical TELRIC on regulatory procedures. It openly invites litigation by econometric models, putting a premium on witnesses who can develop models—typically incorporating only traditionally determined rates of depreciation and return—"demonstrating" that the companies' own incremental costs are far higher than the competitive level.

Note also the regulatory arrogance that it exhibits. Historically, regulators have set rates on the basis of the actual costs of the companies, thus subjecting those rates, as in the majority of states and at the federal level today, to mandated indexations incorporating reasonable expectations of achievable improvements of productivity and, at the same time, opening the market to competition. But instead of leaving it to that process to determine the efficient price, regulators presume to determine those efficient levels and set them at the very outset.

That is not how competition works. Under competition prices tend to be set by the actual costs of the incumbent companies. That provides for entrants both a proper signal and a reward: if they can do better than the incumbent companies, they enter the market and reap the reward of the difference between the rates they are able to charge and the cost-based rates of the incumbent companies. The TELRIC-BS price at which it can instead purchase those facilities from the incumbent simply eliminates both the incentive and the reward.

In short, the sweeping prescriptions of ways the incumbent companies must assist their competitors clearly contradict the central purpose of the Telecommunications Reform Act, which is to encourage genuinely independent competitive entry. Why will any competitor construct its own facilities if it can buy any part of the local network that it needs at a rate explicitly designed to measure the lowest possible cost of the most efficient new competitor that builds a new system from scratch? And why would any competitor offer to produce retail services itself if it can purchase them from the incumbent at a discount prescribed by regulators to cover the total incremental costs of any wholesaler that wants to get into the business?

So once again, by their meddling, under enormous pressure to produce politically attractive results, regulators have violated the most basic tenets of efficient competition—that it should be conducted on the basis of the respective actual incremental costs of the contending parties; and it is that competition, rather than regulatory dictation, that should determine the results. I can think of no clearer vindication of the proposition that whom the gods would destroy, they first make regulators.

2. A People's Constitution

Peter W. Huber

The rhetorical case for dismantling the FCC often begins with the Star Chamber and the First Amendment. Telecommunication is speech. Electronic technology is just a faster printing press, just another star in what Marshall McLuhan called "the Gutenberg Galaxy."[1] The First Amendment covers Gutenberg. It covers Marconi and Bell, too.

But most of this discussion is idle. As it bears on the most serious legal issues of the telecosm, the Constitution has failed. The common law matters more.

State Action

The real constitutional debate ended in 1927, when Congress declared a state of scarcity and nationalized all the airwaves. A few years later Congress declared phone lines to be scarce, too, this time for economic reasons. The lawmakers who announced these facts did not trouble themselves with the Constitution; they created facts instead. No court was going to knock down the only administrative structure that made broadcasting technically possible or phone lines economically viable. No constitutional quibbling was going to stand in the way of immutable scientific reality. None did.

With the FCC itself through the eye of the constitutional needle, the rest was detail. The Commission could constitutionally license electronic presses, since scarcity left no other alternative. The Commission could constitutionally regulate its licensees, because a privileged trustee of government property could naturally be asked to manage the estate as the owner directed. The telecosm was now public, like the Grand Canyon. The new Park Service would regulate

Peter W. Huber is the author of Law and Disorder in Cyberspace *(Oxford University Press, 1997), from which this paper is adapted.*

all vendors of trinkets and postcards. And it would make the premises safe for children, perverts, or sociopaths who might tumble in if park rangers did not maintain the fences.

No one paused to consider just how far this logic might reach. Newspapers use public spaces, too, for trucks and newsracks.[2] Most papers are monopolies as well; few towns have more than one. Happily for newspapers, the Printing Press Commission had been tried in England in 1500 and repudiated on American shores in 1791. But the framers of the Constitution had neglected to mention photons.

So the FCC slipped quietly into place. As the years passed, its constitutional status grew secure. Today few serious lawyers would even dare to suggest that the Commission might be unconstitutional, Title to Title, top to bottom. The big constitutional claim died of neglect in its infancy and has never been seen again.

We have been occupied instead by all the little constitutional issues that swarm out of the Commission like maggots. In 1943 the Supreme Court had to decide whether the FCC could constitutionally regulate network broadcasting.[3] In 1969 the Court had to pass on the constitutionality of the FCC's Fairness Doctrine.[4] In 1984 a federal appellate court had to pass on the constitutionality of licensing direct broadcast satellite (DBS).[5] Half a dozen different courts addressed the constitutionality of rules that barred video programming by phone companies.[6] In 1994 the Supreme Court had to rule on whether Congress could constitutionally require cable operators to carry broadcast signals.[7] Virtually every major licensing decision by the Commission has been attacked on constitutional grounds; those that have not yet will be. Any ruling by the Commission that has any major impact on the structure of the telecosm is constitutionally suspect.

This infestation of constitutional vermin was inevitable. The Constitution limits government's powers, not the people's—state action, not private conduct. A private publisher may set prices, deliver service, provide carriage, and censor content all it likes. It may search, seize, discriminate, and exclude, so long as it does so in a strictly private context. But all of these become constitutional issues when performed to please a commission.

And yet the Commission's rules have mostly been upheld. The reason is almost too discouraging to record. It is like the punch line to the old joke. All we are negotiating now is the price.

The challenge going forward is not to fine-tune the Commission or distinguish good commission law from bad. It is to get beyond the Commission—and thus beyond the Constitution—entirely.

Deconstructing the Telecosm

Get rid of the Commission, and the constitutional debates go away, too. But so does the Commission's power to anesthetize the antitrust laws. Those laws, all in all, protect competition far better than a commission-tolerant Constitution.

Much of the Commission's work, for most of its life, has had the effect of protecting monopoly and promoting scarcity. The Commission separated carriers from content, programmers from transmission facilities, telephone wires from cable, and wires from spectrum. By erecting walls willy-nilly, the Commission saved us from the unlikely threat of one huge monopoly. And it saved us, too, from the far more likely prospect of robust competition.

Why didn't the First Amendment's presumption in favor of a free "marketplace of ideas" get enforced? If the First Amendment means anything, it surely means that a federal agency may not establish and shelter a small group of elite speakers to the exclusion of everyone else. As interpreted by the courts, however, the constitutional standard is flaccid when applied to content-neutral economic laws. The Supreme Court set out the standard in a 1968 case called *United States v. O'Brien*.[8] Laws of that kind pass muster if they serve an "important or substantial" governmental interest unrelated to the suppression of speech, and if they burden expression no more "than is essential" to further that government interest.[9]

These words are weak in the best of times; they are worthless for constraining a large, expert federal agency like the FCC. The Commission is not staffed with fools. Some "important" interest can always be cooked up; it is always possible to postulate some loss of universal service, some threat of monopoly tomorrow, to justify the FCC's own monopoly-protecting walls today.

The antitrust laws take a quite different tack. They pivot on yesterday's schemes to fix prices or carve up territories and on today's market realities. They absolutely forbid all private deals not to compete. If a phone and cable company privately agree not to poach on each other's turf, the dealmakers go to jail. Carving up markets is lawful only when the FCC does it. The Constitution has provided no protection against economic conspiracy by commission. The antitrust laws have provided real protection against conspiracy by private agreement. So long as matters are kept outside the Commission, competition will thrive.

This, then, is the people's constitution for the economics of the telecosm. Not regulation by commission, not even a commission regulated by *O'Brien*. The people's constitution is antitrust law, developed in the courts under the short, broad mandate of the Sherman Act. The right role—the only constitutionally legitimate role—for Congress and its agencies is to establish property rights, dezone bandwidth, open entry, and then let the market be. Antitrust laws will take care of the rest.

Making Connections

In 1776 Pennsylvania's Constitution declared that printing presses should be common carriers. The presses, it said, were to "be free to every person who undertakes to examine the proceedings of the legislature, or any part of government."[10] The intensely partisan colonial newspapers were thus directed to open their columns to opposing viewpoints and carry all contending opinions, much like the mails. According to contemporary left-wing theory, the First Amendment not only permits but requires much the same today.[11]

In 1939 the Supreme Court itself gave a passing nod of approval to something along those lines. The streets and parks, Justice Roberts observed, "have immemorially been held in trust for the use of the public and, time out of mind, have been used for purposes of assembly, communicating thoughts between citizens, and discussing public questions."[12] This stray comment has evolved into something called "public forum doctrine." What it dimly suggests is a body of common carriage law emerging from the shadows and penumbras of the First Amendment.

Today's streets and parks are in the airwaves, which the federal government has nationalized, or in glass and wire, which run under public streets controlled mostly by local authorities. Is the government obliged, then, to maintain a Public Forum Commission to see to it that diverse voices are heard over these inescapably public media?

Public forum doctrine is a stopgap, a last line of defense after government has taken control of all the real estate and there is nowhere left to speak but on public land. Trying to pluck common carriage obligations out of the obscure depths of the First Amendment, however, does more harm than good. The First Amendment itself implies negative rights, too—rights not to salute, not to pledge,[13] not to embrace government propaganda, not to publish replies, not to insert proclamations in bills, not to disclose membership lists, not to display political slogans on license plates, and not to pay unwillingly for the political speech of others.[14] Rights like those cannot be squared with mandates that require cable companies to open their networks to pornographers or that require TV stations to air commercials denouncing commerce.

Price regulation and universal service are constitutionally indefensible, too. No Federal Newspaper Commission would be permitted to regulate the price of the *Washington Post* or demand that it be sold at uniform prices on every corner of the city.[15] Any requirement of price-averaged, universal service puts the provider and all its customers in a single economic pot. Your phone bills subsidize or are subsidized by escort services, handgun dealers, and dial-a-porn operators. The telephone company must put up with the whole lot. Neither association nor speech is anywhere close to free when economic cohabitation of that order is required by five commissioners in Washington.

Even more plainly unconstitutional are carrierlike duties imposed by commission to promote fairness in the marketplace of ideas. The FCC enforced the Fairness Doctrine against radio and TV broadcasters for years. The Supreme Court upheld the rule in 1969, in a case pitting Lyndon Johnson's FCC against a radio station's defense of Barry Goldwater.[16] In 1974, by contrast, the Supreme Court struck down a Florida law that required any newspaper attacking the record of a political candidate to offer an opportunity to reply. "A newspaper is more than a passive receptacle or conduit for news, comment, and advertising," the Court reasoned.[17] To its shame, the Court failed even to mention, still less distinguish, its Fairness Doctrine precedent.

Any Public Forum Commission would also collide with the Takings Clause of the Fifth Amendment. Public forum doctrine affirms the public's right to speak out on government property; takings law affirms the private owner's right to maintain peace and quiet on his own.[18] Private property includes bandwidth, compression algorithms, and databases, all of which reside in glass or silicon. When government seizes such things, it is seizing property, no matter how worthy its ultimate motives. Micro-takings are no different than macro ones. No one doubts that the Fourth Amendment's search-and-seizure clause applies in the telecosm. The Fifth Amendment's Takings Clause should, too.

So, do all common carriers operate under a body of law that is simply unconstitutional? Yes, so far as common carrier law is imposed unilaterally by Commission edict. But traditional common carrier principles do not originate in Commission mandates. Individual carriers freely elect common carrier status in exchange for limits on their own liability. The common carriers of the common law assume common-carrier status voluntarily. There is nothing unconstitutional about that.

This, then, is the people's constitution for connection to the telecosm. No Pricing Tribunal, Public Forum Commission, or Fairness Doctrine. Just the common law of common carriage, engaged by choice, not by compulsion.

The Marketplace of Ideas

So much for economic matters. Beyond lie the great issues of the information age: free speech and its complements, property and privacy. Ironically, the Constitution is all but irrelevant here. The First Amendment is, of course, of some general interest. But so is Article I, section 8, of the Constitution, which empowers Congress to enact a copyright law. And so too is the Fourth Amendment, which addresses searches, seizures, and, by implication, privacy.

The whole purpose of the second and third rights is to limit the first. In the reign of Queen Mary copyright law was an instrument of censorship. The old Soviet government sought to manipulate international copyright law to suppress books worldwide.[19] The whole point and purpose of copyright is to limit free speech.[20] Privacy rights limit it, too.[21] Electronic privacy is the right not to be put unwillingly at the wrong end of someone else's telecommunicating machine. Every right to speak collides with some reciprocal right not to listen, not to be heard, not to see or be seen, not to surrender one's own words, thoughts, privacy, or solitude.

Commissions try, of course, to sort out conflicts such as these. They give a bit of free speech here, for political protest, a bit of copyright there, for cablecasters, and a little privacy at the next turn, for people who might otherwise be harassed by telemarketers.

With the crude, indiscriminate media of over-the-air broadcast—electronic sound trucks, so to speak—collectivizing choices like these seemed inevitable. Collectivizing the media collectivized the message. But the precise media of narrowcasting offer far better private substitutes. By arming individuals, technology makes it possible to disarm government. The key to truly free speech—mutual consent—can be vindicated without government interference or assistance. Indeed, it cannot be vindicated any other way.

Technology is at hand to protect the negative dimensions of free speech, copyright, and privacy. Caller ID, backed by computer screening, offers a new level of privacy from and control of the assault by telephone. The V-chip will empower parents to limit what their children watch, and it will soon be succeeded by more powerful chips that permit precisely calibrated private censorship. Communities of speakers and listeners then become completely consensual. The logic for management by commission disappears.

A robust copyright law promotes speech in much the same way. Copyright is what gives people a real, tangible, financial stake in speech itself. And it is that stake, not the dry words in the Constitution, that supplies the most durable protection of a free press. Despite its disreputable origins three centuries ago, copyright was the "uniquely legitimate offspring of censorship."[22]

There remains speech that inspires psychotics, or motivates sociopaths, or defames reputations. But here too the main lines of defense can be private, not public. Liability suits can deter communication that incites violence—and such suits can properly take aim at conduct, not speech itself. Private libel suits ensure fairness and honesty much better than a fairness doctrine.

The final objection to brokering those rights in a commission is the most fundamental. Collectivizing individual rights eviscerates them. The whole point of copyright is to privatize what is otherwise too public. Move copyright out of Steven Spielberg's hands and into a Copyright Royalty Commission's, and Spielberg's rights are sharply devalued. Move privacy itself out of private hands, and a commission will simply end up splitting the difference between people who prefer more solitude and people who prefer less. The very process of transferring rights such as these from private to public hands defeats them. The rights exist only so long as they are exercised by one individual at a time, not by a commission on behalf of all.

This, then, is the people's constitution for the marketplace of ideas: private property and private choice, privately exercised in open markets. And private right, privately enforced, in common law courts.

Common Law for the Telecosm

The legal debate ultimately pivots on two profoundly different engines for the making of law. One is top-down: it is law by edict and national commission. The other is bottom-up: law built by adjudication in common law courts. One kind of law occupies thousands of pages in the U.S. Code and the *Federal Register*. The other evolves as a pure product of common law, or under short, general mandates like the Bill of Rights or the Sherman Act.

In the telecosm, as elsewhere, commission law leads society down the road to serfdom.[23] However good the original intentions, central planning always ends up maintaining the privilege and power of the planners themselves. From markets and the common law, by contrast, there emerges spontaneous order that is rational, efficient, and intelligent. Though never planned, never even fully articulated, common law rules adapt and evolve by common consent, like the rules of grammar. Society organized by commission is inherently limited by what the minds of the planners can grasp. Common law, in the aggregate, is far wiser. England, the land of the Magna Carta, never did get a written constitution. But it got the common law, and with it the most stable, decent, and consensual legal order on the planet.

Small-scale and privately centered common law is the only kind of law that sits comfortably with our traditions of individual freedom and private liberty. The first word of the First Amendment is "Congress," and the phrase that follows is "shall make no law." In the arena of communication, the federal government is the principal suspect, the least trusted branch. The bias is against the big, not the small; against state action, not private undertaking; against the commission, not the private citizen.

Nothing grander than common law is even practical anymore. The telecosm is too large, too heterogeneous, too turbulent, too creatively chaotic to be governed wholesale, from the top down. That is, of course, unsettling. The stability of management from the top is much more reassuring. But only because a commission's greatest power is to maintain the status quo. The telecosm's promise is to transform it.

Left to common law, the telecosm will become again a place of vast freedom and abundance. There will be room enough for every sight and sound, every thought and expression that any human mind will ever wish to convey. It will be a place where young minds can wander in adventurous, irresponsible, ungenteel ways.[24] It will contain not innocence but a sort of native gaiety, a buoyant, carefree feeling, filled with confidence in the future and an unquenchable sense of freedom and opportunity.[25]

Notes

1. Marshall McLuhan, *The Gutenberg Galaxy: The Making of the Typographic Man* (Toronto: University of Toronto Press, 1962).
2. *See Miami Herald Publishing Co. v. Tornillo*, 418 U.S. 241 (1974); *City of Lakewood v. Plain Dealer Publishing Co.*, 486 U.S. 750 (1988).
3. *National Broadcasting Co. v. United States*, 319 U.S. 190 (1943).
4. *Red Lion Broadcasting Co. v. FCC*, 395 U.S. 367 (1969).
5. *National Ass'n of Broadcasters v. FCC*, 740 F.2d 1190 (D.C. Cir. 1984).
6. *See Chesapeake & Potomac Tel. Co. v. United States*, 42 F.3d 181 (4th Cir. 1994), *aff'g Chesapeake & Potomac Tel. Co. v. United States*, 830 F. Supp. 909, (E.D. Va. 1994); *U S West, Inc. v. United States*, 1994 U.S. App. LEXIS 39121 (9th Cir. 1994), *aff'g U S West, Inc. v. United States*, 855 F. Supp. 1184 (W.D. Wash. 1994); *Southwestern Bell Corp. v. United States*, No. 3:94-CV-0193-D (N.D. Tex. Mar. 27, 1995); *USTA v. United States*, No. 1:94CV01961 (D.D.C. Jan. 27, 1995); *GTE South, Inc. v. United States*, No. 94-1588-A (E.D. Va. Jan. 13, 1995); *NYNEX Corp. v. United States*, No. 92-323-P-C (D. Me. Dec. 8, 1994); *BellSouth Corp. v. United States*, 868 F. Supp. 1335 (N.D. Ala. 1994); *Ameritech Corp. v. United States*, 867 F. Supp. 721 (N.D. Ill. 1994).
7. *Turner Broadcasting System, Inc. v. FCC*, 114 S. Ct. 2445 (1994), *reh'g denied*, 115 S. Ct. 30 (1994).
8. *United States v. O'Brien*, 391 U.S. 367 (1968).
9. Ibid., at 377.
10. The Constitution of Pennsylvania, September 28, 1776, ¶38, reprinted in S. E. Morison, *Sources and Documents Illustrating the American Revolution, 1764–1788, and the Formation of the Federal Constitution*, 2d ed. (Oxford: Clarendon Press, 1929), pp. 162, 173.
11. *See*, e.g., Jerome A. Barron, *Freedom of the Press for Whom? The Right of Access to Mass Media* (Bloomington, Ind.: Indiana Univ. Press, 1973); Jerome A. Barron, "Access to the Press—A New First Amendment Right," Harvard Law Review 80 (1967): 1641.
12. *Hague v. Committee for Indus. Org.*, 307 U.S. 496, 515 (1939).
13. *See West Virginia Board of Education v. Barnette*, 319 U.S. 624 (1943).

14. *See* generally Laurence H. Tribe, *American Constitutional Law*, (Mineola, N.Y.: Foundation Press, 1978), § 12.4, p. 589.

15. Cable plaintiffs have launched a frontal attack on the rate-regulating provisions of the 1992 Cable Act. *Daniels Cablevision, Inc. v. United States*, 835 F. Supp. 1, 7 (D.D.C. 1993), vacated and remanded sub nom., *Turner Broadcasting Sys. v. FCC*, 114 S. Ct. 2445 (1994), *claim dismissed, summ. judgment granted*, 910 F. Supp. 734 (D.D.C. 1995), *vacated and remanded sub nom., National Interfaith Cable v. FCC*, 114 S. Ct. 2730 (1994). See also Time *Warner Entertainment Co., L.P. v. FCC*, 810 F. Supp. 1302 (D.D.C. 1992); *Time Warner Entertainment Co., L.P. v. FCC*, 1996 U.S. App. LEXIS 22387 (D.C. Cir. Aug. 30, 1996).

16. *Red Lion Broadcasting Co. v. FCC*, 395 U.S. 367 (1969).

17. *Miami Herald Publishing Co. v. Tornillo*, 418 U.S. 241, 258 (1974).

18. In 1980 the Supreme Court upheld a California state law permitting solicitation of signatures in the courtyard of a privately owned mall, but only because the value and use of the shopping center had not been measurably impaired. *Pruneyard Shopping Ctr. v. Robins*, 447 U.S. 74 (1980). Telephone companies won a 1994 takings challenge to an FCC order that required them to permit physical collocation of competitors' wires and switches on telephone company premises. *Bell Atlantic Tel. Co. v. FCC*, 24 F.3d 1441 (D.C. Cir. 1994).

19. The transfer of rights under copyright may be "by any means of conveyance," including a will, or "by operation of law." 17 U.S.C.A. §201(d)(1). This does not include, however, government seizure from an unwilling author. 17 U.S.C.A. §201(e).

20. By its terms, the Copyright Act is intended not to displace the First Amendment. Pub. L. No. 650, Tit. VI, §609, 104 Stat. 5132 (1990). The Supreme Court has addressed the tension between the First Amendment and copyright in a number of cases. See e.g., *Mazer v. Stein*, 347 U.S. 201, 217–18 (1954).

21. *See* Tribe, p. 796.

22. Paul Goldstein, "Copyright and the First Amendment," *Columbia Law Review* 70 (1970): 983.

23. Friedrich Hayek, *The Road to Serfdom* (Chicago: University of Chicago Press, 1943).

24. George Orwell, "Review, *Herman Melville*, by Lewis Mumford," in *The Collected Essays, Journalism, and Letters of George Orwell*, ed. Sonia Orwell and Ian Angus (New York: Harcourt, Brace & World, 1968), p. 21.

25. George Orwell, "Riding Down from Bangor," in *The Penguin Essays of George Orwell* (1984), pp. 406, 407.

3. The Frontier of Telecommunications Deregulation: Small Countries Leading the Pack

Pablo T. Spiller and Carlo G. Cardilli

Four small countries—Australia, Chile, Guatemala, and New Zealand—are at the forefront of telecommunications deregulation. Chile's 1982 General Law on Telecommunications was the first; like many pioneers, Chile learned some lessons the hard way. But today Chile has the fiercest long-distance and local telephone competition in the world. New Zealand tried to jump straight from regulation to unrestricted competition on April 1, 1989. The Australian government learned from New Zealand's experience with its 1991 Telecommunications Act. Finally, Guatemala enacted sweeping reforms in November 1996, learning from the experiences of Chile, New Zealand, and Australia, as well as the slower progress of telecommunications deregulation in the United States, the United Kingdom, and Mexico.

As each country has learned from the others, there are many common elements in these countries' reforms. In each country a politically strong government was committed to radical reforms. The incumbents were all state-controlled monopolies. In some respects, the countries are different from the United States, yet they have much to teach U.S. lawmakers and regulators.

Interconnection, equal access, unbundling, and industry structure are the four building blocks that determine how quickly facilities-based competition will emerge once the telecom sector is deregulated. We describe key decisions affecting how the four countries employed the building blocks and assess the effect on competition

Pablo T. Spiller is a professor at The Haas School of Business in Berkeley, California. Carlo G. Cardilli is managing economist at LECG, Inc. *A more detailed version of the paper appears in the* Journal of Economic Perspectives 11, no. 4 (Fall 1997): 127–138. Reprinted with permission.

and consumer welfare. We find that market mechanisms are superior to regulatory processes. But market mechanisms need clear rules and enforcement mechanisms (such as final-offer arbitration) that do not invite manipulation.

Implementing the Building Blocks of Competition

Interconnection

Interconnection is the right of a network operator to ensure that its users can make or receive calls to or from the users of another network. Telephony incumbents often have denied interconnection to new entrants. Without a right to interconnection, the incumbent can combine network externalities with its installed base to foreclose other competitors.[1] But given the right to interconnect, the local telecommunications market is not a natural monopoly, as the average costs of serving a local exchange area do not increase appreciably once a small minimum efficient size is reached. Almost the only important economy of scale in local exchange service is that of interconnection.[2]

The network externality makes incumbent monopolies likely to deny, delay, or overprice interconnection to preserve their dominant position, thus requiring entrants and small competitors to actively enforce their interconnection rights. As a consequence, the mechanism by which the interconnection right is enforced is the most important influence on facilities-based competition after the creation of the right itself. The right to interconnection was neglected in both Chile and New Zealand, with serious adverse consequences.

Before deregulation in Chile, CTC provided most local telephony, while ENTEL provided most long-distance services, and both were state-controlled corporations. The 1982 General Law on Telecommunications mandated open and nondiscriminatory access to the incumbent networks for private competitors and deregulated prices but failed to specify interconnection procedures or pricing. Years of interconnection-related lawsuits resulted, imposing heavy losses on entrants.[3] Eventually, a second set of reforms required entrants and incumbents to resolve disputes by arbitration and gave the Chilean regulator, SUBTEL, substantial powers. SUBTEL set interconnection charges based on long-run average incremental costs. Local competition took hold, with the CMET and Manquehue telephone companies beginning to overbuild CTC's network in Santiago and surrounding areas. Chile replayed this scenario with long-distance competition, when ENTEL's long-distance monopoly was abolished in 1992; again, vague rules engendered acrimonious litigation until SUBTEL stepped in.

A similar problem arose in New Zealand after the privatization of the incumbent (Telecom Corporation of New Zealand-TCNZ) in 1990. Trying to avoid overregulation, the New Zealand reformers relied solely on generic antitrust law to check the incumbent's monopoly power.[4] A torrent of litigation followed.

The first long-distance interconnection was concluded between Clear Communications (the entrant) and TCNZ on March 4, 1991, almost two full years after deregulation. Attempting to speed progress, the Commerce Commission (the agency charged with antitrust enforcement) started an inquiry in November 1991, which TCNZ halted through litigation. More litigation erupted in 1991 over the payment from Clear to TCNZ for the interconnection necessary to provide local service and was resolved after years of expensive litigation by the Queen's Privy Council in London. The New Zealand government, dissatisfied with both the length and the outcome of the court battle (the outcome was efficient-component pricing),[5] threatened to regulate if the parties could not reach a satisfactory agreement.[6] Only six weeks later TCNZ and Clear announced agreement on many issues.[7] The pace of interconnection appears to have sped up of late. Both Sprint and Telstra recently signed interconnection contracts with TCNZ, the Telstra deal reportedly requiring only eight meetings over a four-month span.

Australia, apparently taking notice of events in New Zealand, explicitly established the right to interconnection in the 1991 Telecommunications Act. The government instructed AUSTEL, the regulator, to set interconnection rates based on "directly attributable incremental costs." The rates were to be transitory, applying only to traffic originating or terminating within the area where the interconnecting party maintained a point of interconnection, and only until the incumbent was dominant in that area (which typically means that the entrant has less than a 20 percent share of the market in that area). AUSTEL was designated as the arbitrator in cases of dispute over interconnection agreements. AUSTEL quickly steered parties in the direction of negotiated agreements; as of 1995, six of seven agreements did not require AUSTEL arbitration.

The 1996 law deregulating telecommunications in Guatemala also established a fundamental right to cost-based interconnection for all telecommunications carriers. To avoid protracted litigation, the Guatemalan legislature opted for negotiation under a specific timetable (as short as four months), requiring any disputes to be resolved through final offer arbitration administered by a specific industry regulator.

Final-offer arbitration (also known as pendulum arbitration or as the baseball rule) is a dispute resolution mechanism that limits the parties' posturing incentives. Its key feature is that the arbitrator chooses only between the two final offers presented by the disputing sides, and the chosen offer then becomes binding on both parties. The arbitrator therefore does not need to fashion an intermediate or compromise solution, and parties have an incentive for truthful revelation. Either party gains by making its offer marginally fairer than the opponent's expected offer, giving both parties a strong incentive to make reasonable offers and reach a negotiated agreement rather than resort to arbitration.[8] In fact, if the arbitration is time-constrained, as it is in Guatemala, the incumbent has little to gain by procrastination, and the dominant strategy is typically to reach agreement with the entrant and get down to the real business of competition.

The key to the Guatemalan approach was that the legislation did not specify how agreements should be framed; instead it explicitly listed the services that network operators must offer to rivals and specified explicitly the cost-based rules that the arbitrator must follow to select the best offer from the disputing parties. Consequently, interconnection took place speedily, and the first interconnection agreement was concluded in August 1997, three months after the interconnection right came into effect. Later in 1997, and under pressure to privatize the government telecommunications operator, the government introduced legislation substantially altering the cost-based nature of the arbitration rule.

In summary, the crucial issue is the method chosen to prevent the incumbent from creating a negotiating deadlock. The experiences of Chile and New Zealand show that the use of courts to resolve disputes in the telecommunications arena has been slow and unpredictable. Relying only on antitrust law as the backstop to a negotiated interconnection process does not seem to work well. Australia's experience shows that a traditional industry-specific regulator can be successful; but, in general, a pure regulatory approach such as Australia's concentrates the decisionmaking in one set of hands and gives parties the (wrong) incentive to make exaggerated and continuous claims in hopes of extracting favorable treatment.[9] Guatemala has successfully used binding arbitration with specific resolution criteria known ex ante as a backstop to encourage negotiated settlements.

Equal Access

Equal access is a system that permits consumers to select the carrier that will deliver calls to their final destination by dialing the same number of digits irrespective of the identity of the carrier. Thus, a system that allows the customer to select a default carrier and requires an access code for all other carriers constitutes equal access. A system where the calls default to the provider of local exchange service unless additional digits are dialed is not.

In New Zealand, Australia, and Chile, the deregulating legislation did not specify equal access, but each country pushed toward its own version. In New Zealand the threat of antitrust litigation led TCNZ to agree to provide equal access to Clear, first giving TCNZ subscribers access to Clear through the special access code 050, and equal access once Clear's share reached 9 percent. But TCNZ stretched out the implementation of equal access by requiring Clear to slog through years of mediation, lawsuits, and arbitration. In Australia, equal access was demanded by the regulator and implemented gradually.

In contrast to the gradual implementation of equal access in Australia and New Zealand, the Chilean implementation was explosive. A supreme court decision forced Chilean regulators to take a radical approach. Regulators thus mandated that all long-distance carriers be accessible only by dialing the carrier's unique access code, a requirement that forced customers to select a specific long-distance carrier every time they made a call. The entire country was given equal access virtually overnight. Prices of long distance fell dramatically, and consumption increased. Guatemala sought to emulate Chile's success, and thus recently mandated that all telecommunications operators above a minimum size provide nondiscriminatory equal access to all other interconnected operators.

Unbundling

Unbundling requires the incumbent to offer entrants basic network elements and services, such as lines, switching, and transport, individually. An entrant can lease a few assets from the incumbent, an arrangement that increases the prospect of successful entry. For example, a local carrier might buy access to unbundled local lines and hook them up to its own switching system, or a cable company might purchase access to unbundled ports and switching but use its own lines to provide dial tone to its customers. Unbundling requirements, however, can stifle innovation and investment, especially if the incumbent must partition its services too finely. Such policies risk causing entrants to focus on exploiting inefficiencies in the regulated unbundled rates and pick incumbent networks apart, instead of investing in the industry.

Australia, Chile, and New Zealand did not require their incumbents to unbundle their services extensively. But the Guatemalan government wanted instant competition and used unbundling as a transitory measure. In exchange for fully deregulating end-user rates, carriers in Guatemala are required to unbundle four components of their networks—loops, ports, switching, and signaling functions—at cost-based rates. To discourage pure resellers, the obligation to unbundle is "coarse" and explicitly transitory (expiring in 1999) and does not require the incumbent to provide transport. Thus, any entrant in Guatemala has to make some investment in wholly owned facilities.

Industry Structure

Since most deregulating countries have started off with a vertically integrated monopoly, the first choice to be made is whether to break it up or not. The second important choice is the extent of competition to be introduced into the system.

Australia, New Zealand, and Guatemala all chose to keep the incumbent vertically integrated but differed greatly in the number of competitors they were willing to allow. Australia (following the lead of the United Kingdom) authorized only one full-fledged competitor to the incumbent in wireline and two in wireless between 1991 and 1997. Resellers were permitted in Australia, but they depend on the goodwill of the incumbent for interconnection or equal access.

By contrast, both New Zealand and Guatemala exposed their incumbents to unfettered entry of any kind right from the start.

Chile, on the other hand, began its reforms with structural separation already in place: one state-controlled company dominated local telephone service while another state-controlled firm dominated long-distance service, each facing limited competition from smaller private firms. In 1994 CTC and ENTEL were allowed to enter each other's business—with some market share restrictions—while all restrictions on entry by others were lifted.

The Effects of Deregulation on the Telecommunications Market

Facilities-based competition has arrived in all these countries in every market segment—nowhere more than in Chile. Deregulation has had striking effects on the price and quantity of service in all four nations. The intensity of competition in local service has been especially intriguing, particularly as it was long considered (mistakenly, in our view) a natural monopoly.

In New Zealand, the incumbent TCNZ has now concluded interconnection contracts with five competitors and competes with them in both domestic and international toll markets. Clear, the largest competitor, has around 23 percent of the domestic and international toll markets and attempted to enter local telephony less than two years after deregulation. Cable TV providers are building mixed telephony/cable networks in metropolitan areas, as has happened in the United Kingdom.[10] Interconnection rates are low by international standards and compare favorably with rates in the United States for similar services. TCNZ's average toll rates declined 31 percent in real terms between mid-1991 and mid-1996,[11] while the overall price of the basket of residential telecommunications services declined at an annual rate of 2.5 percent over the same period.[12] Local exchange rates for large businesses in major central business districts have also declined 15 percent since 1991.[13] Local rates have continued to drop in the aftermath of a toll war that started in September 1996, spurred by the new popularity of cellular service.[14]

Telecommunications competition in Australia was somewhat handicapped by the duopoly decision on industry structure that led to rather tame competition. The only permitted wireline entrant did not introduce its local exchange product until 1996, almost five years after deregulation.[15] Predictably, Telstra's prices for local exchange services fell little more (1 percent) than required by its price cap.[16] However, duopoly competition led to a slight decrease in Telstra's domestic long-distance rates of 5.5 percent and 9.0 percent in 1993–94 and 1994–95, respectively. The single entrant performed somewhat better in international long distance, capturing about 21 percent of outbound minutes by the end of 1994. Correspondingly, Telstra's international rates decreased 8.7 percent and 7.6 percent in 1993–94 and 1994–95, respectively.[17] Telstra's international share has recently fallen below 60 percent.[18] Despite the requirement that customers dial a four-digit code to access their services, resellers have been successful as well. The largest reseller provides long-distance service to more than 450,000 residential and business access lines (about 5 percent of the Australian total), as well as bulk bandwidth for Internet service providers, mostly over its own infrastructure that links the major cities.

In Chile, eight companies are now providing long-distance service, and SUBTEL has approved three more to begin service soon. Prices have plummeted; by September 1996, average long-distance rates had fallen by more than 50 percent since the liberalization of long distance in 1994, and domestic long-distance traffic had risen by more than 50 percent. Crandall and Waverman estimate that the reforms unlocked approximately $120 million (U.S. dollars) in consumer welfare in 1995 alone. To survive the rugged competition, almost all long-distance companies have adopted a strategy of becoming integrated providers, concentrating on a particular set of customers rather than a set of services. Nearly all carriers plan to enter the local loop.

Because Chile has no unbundling or resale requirements, construction of facilities is the only way to provide local service. As a consequence, there currently are six actively competing local exchange providers in and around Santiago, including the previously state-controlled companies CTC and ENTEL; two original competitive local exchange carriers in Santiago, CMET and Manquehue; VTR, a large cable-TV provider; and a Telefónica Andina, a new entrant. In many parts of Santiago consumers already have a choice of two or three local providers, with more to come. In May 1996, the incumbent's share of local access lines in Santiago was expected to dip below 75 percent by the end of 1997.[19]

Competition in Guatemala started immediately once free entry was permitted in November 1996, despite the fact that the right of interconnection to the incumbent was not immediately available. As of July 1997 there were 160 registered telecommunications operators in the country, many providing international satellite services using informal line-side interconnection (i.e., interconnecting just as regular telephone subscribers do). Internet service provision expanded just as fast, from 1 provider to currently more than 30. The penetration of cellular telephony also has increased remarkably; it now accounts for more than one-eighth of the access lines currently in service.

A comparison of rates in the early reforming countries—Chile, New Zealand, and Australia—shows that competition has driven prices lowest in Chile. Australia has probably been held back by its duopoly policy, while competition in New Zealand has probably been held back by the long-running interconnection and equal-access litigation. Guatemala has done disproportionately well, for mainly historical reasons—the Guatemalan government's adoption of price reform and international rate rebalancing in 1996.

A Summary from the Frontier

Policy decisions concerning telecommunications deregulation do affect competitive outcomes. Loosely specified interconnection rules lead to delays, as in Chile and New Zealand. Ignoring the issue of equal access puts entrants at a clear competitive disadvantage, while the evidence from Chile shows the substantial gains to be had by introducing equal access with a requirement that a long-distance carrier be explicitly selected for every call. Some evidence on the merits of temporary and limited unbundling requirements to stimulate an initial surge of competition will arrive in the next few years from Guatemala. However, Chile's absence of unbundling requirements, combined with the players' strong tendency toward vertical integration (which suggests strong economies of scope at work), has helped unleash a remarkable level of competition in the provision of local services. Finally, restricting the number of entrants only delays the emergence of true competition and does not lead to "stronger" competitors.

The experience of the four countries readily disproves the myth that telecommunications remains a natural monopoly where competition must be engineered by regulators. Indeed, facilities-based telecommunications competition is a hardy plant that can prosper in many different regulatory environments, often with startling gains.

Notes

We are indebted to Tim Taylor for many editorial suggestions and to Alan Krueger and Brad DeLong for incisive comments. We also thank participants at the workshops sponsored by the Inter-American Development Bank in El Salvador (Spring 1996) and by the U.S. Agency for International Development in Guatemala (Summer 1996) and Costa Rica (Spring 1997), where we presented many of the concepts in this article, and to Heather Noss for her excellent research assistance.

1. Nicholas Economides, "The Economics of Networks," *International Journal of Industrial Organization*. 14, no. 6 (October 1996): 673–99.

2. Jerry A. Hausmann and Timothy Tardiff, "Efficient Local Exchange Competition," *The Antitrust Bulletin* (Fall 1995): 540–42.

3. Renato Agurto and Jorge F. Asecio, "Telecommunications and Transport," in *Private Solutions to Public Problems: The Chilean Experience*, ed. Cristian V. Larroulet (Santiago, Chile: Instituto Libertad y Desarrollo, 1993), p. 237.

4. See Patrick G. McCabe, "New Zealand: The Unique Experiment in Deregulation," in *Telecommunications in the Pacific Basin: An Evolutionary Approach*, ed. Eli Noam, Seisuke Komatsuzaki, and Douglas A. Conn (New York: Oxford University Press, 1994), p. 410; David Galt, "Telecommunications Regulatory Structures in New Zealand," Presentation to International Telecommunications Society's Interconnection Workshop, Wellington, April 10, 1995.

5. Efficient-component pricing is explained in William J. Baumol and J. Gregory Sidak, "The Pricing of Inputs Sold to Competitors," *Yale Journal of Regulation* 11, no. 1 (1994): 171–202.

6. New Zealand Ministry of Commerce and New Zealand Treasury, "Regulation of Access to Vertically-Integrated Natural Monopolies," Auckland, 1995.

7. Peter Gorringe, "The Regulation of Telecommunications in New Zealand," mimeo, 1996.

8. See, for example, B. J. Coleman, K. M. Jennings, and F. S. McLaughlin, "Convergence or Divergence in Final-Offer Arbitration in Professional Baseball," *Industrial Relations* 32 (1993): 238–47.

9. Pablo T. Spiller, "A Contract Test Approach for Negotiated Interconnection: Implementing the Telecommunications Act of 1996 to Foster Competition in Local and Long Distance Services," in comments of Bell Atlantic in FCC Common Carrier Docket no. 96-98.

10. Rt. Hon. Maurice Williamson, "Telecommunications—New Zealand's Competitive Edge," Transcript of speech given at the New Zealand Telecommunications Summit, Wellington, 1996.

11. Peter Gorringe, "The Regulation of Telecommunications in New Zealand," mimeo, 1996; David B. De Boer and Lewis Evans, "The Economic Efficiency of Telecommunications in New Zealand," *Economic Record* 72 no. 216 (1996): 24–35.

12. New Zealand Ministry of Commerce and New Zealand Treasury, "Regulation of Access to Vertically-Integrated Natural Monopolies," Auckland, 1995.

13. Williamson.

14. "Telecom NZ and Clear in Price War," *Exchange* 8 no. 36 (September 20, 1996): 3.

15. "Telstra About to Be Unleashed," *Melbourne Age*, March 31, 1997, p. B1.

16. AUSTEL, *Australian Telecommunications Authority 1994–5 Annual Report* (Canberra: Australian Government Publishing Service, 1995).

17. Ibid.

18. "Telstra Facing Revenue Crisis," *Melbourne Age*, Australia, February 3, 1997, p. C1.

19. "Competencia en Telefonía Local: El Próximo Frente," *Estrategia* (Santiago, Chile), May 22, 1996.

Part II

Interconnection and Infrastructure in a Free Market

4. Fulfilling the Intent of Congress under the Telecommunications Act of 1996

Tom Tauke

The principal purpose of the Telecommunications Act is to open all telecommunications markets—local, long distance, video, and wireless—to competition. While pundits often focus on the local market, Congress recognized that long-distance markets are less competitive and innovative than they could be. It believed the Regional Bell Operating Companies (RBOCs) could offer strong competition in long distance and wanted to ensure that they could soon enter the market.

Congress wisely determined to use the incentives of the marketplace to achieve fully open and competitive markets. It relied principally on private negotiations between parties to interconnect the networks of long-distance companies, competitive local exchange companies, cable companies, and wireless firms.

Congress recognized that communications markets are changing. Customers increasingly want convenience, value, and a variety of services—from video to Internet to local service—marketed in ways that meet their often unique needs. The way we think of communications services—such as local and long distance—is changing, and marketing such services in the traditional manner, service by service, is not going to meet the needs of many consumers.

In such a market, all companies will have to find new service mixes that "hit the mark" with today's diverse consumer segments, creating new value through the convenience of one bill. Long-distance companies will have to include local services in their marketing efforts; local companies will have to be able to offer long distance; and other companies, such as cable firms, will need to

Tom Tauke is the senior vice president for government relations with Bell Atlantic Corporation.

create new communications options too. All firms will have to be able to access all markets.

In this environment, Congress believed that long-distance firms, cable companies, and local exchange providers would have incentives to negotiate the interconnection of their networks, thus ensuring rapid entry into all markets.

Evidence That the Telecommunications Act Is Working

The good news is that the local competition aspects of the Telecommunications Act are working. Nationally, to date, the incumbent local exchange companies have signed over 1,200 interconnection agreements with competitors. Bell Atlantic has entered into more than 605 interconnection agreements since the passage of the Telecommunications Act. Some 120 of those agreements are with facilities-based wireline local competitors, and 140 have already been approved by the state regulators.

Bell Atlantic has also been preparing in other ways for local competition. The company has completed work on over 530 collocation sites, and another 417 sites are in the process of completion. More than 3,000 telephone exchange codes, each serving thousands of individual telephone numbers, have been allocated to competitive local exchange companies for their use. Bell Atlantic has installed over 370,000 trunks to exchange calls with competitive local exchange carriers. We estimate that competitors now serve more than 1 million customer lines in local markets throughout our region.

But local competition is only part of the story. The IntraLATA or regional toll market—long-distance calls that go outside a local exchange but generally do not cross state boundaries—is competitive and open to entry in all of Bell Atlantic's territory, and in most states across the United States. This is a lucrative market, worth some $20 billion nationally, that until recently most long-distance companies, such as MCI, AT&T, and Sprint, were precluded from entering. It was regulated and the province of the incumbent local exchange companies. Today, the IntraLATA is a very competitive market, one that any company may enter with relative ease.

The Internet itself is helping to create a vibrant market that was once of concern only to businesses—the data market. Some 60 million Americans log on to the Internet on a regular basis and 4,300 Internet service providers (ISPs) are servicing this market. In Bell

Atlantic's territory, data traffic levels will soon equal levels for Inter-LATA or long-distance traffic. The Internet is becoming integrated with local networks and its growth is even leading to congestion on the voice-engineered local telephone system. Cable companies—through cable modem technology—and wireless providers are also beginning to provide local access to the Internet in competition with the incumbent companies.

The growth of the Internet is a welcome and positive development to Bell Atlantic, but the Net is both a source of growth and a new form of competition. E-mail exchanges can substitute for what might once have been done over the voice network. While still small, Internet telephony is a growing market, and faxing over the Internet is already a major factor. In fact, estimates are that today some 30 percent of all business faxing is already done over the Internet. Major companies have announced Internet faxing services that are very attractive to business customers. Bell Atlantic recognizes the competitive importance of the Internet and has introduced its own Internet access service.

Why Some Perceive the Act Has Failed

If the act is beginning to work and competition is entering the marketplace, why are there some who allege that the act is a failure, that the local exchange companies are impeding competition, and that the regulators—the Federal Communications Commission and the state Public Service Commissions—should step in to set things right?

Impatience and unrealistic expectations explain some of the criticism. In the effort to gain passage of the Telecommunications Act last year, the promise of what competition could bring to consumers and the country sometimes obscured the amount of work and time that would be needed to lay the groundwork for local competition. After all, long-distance competition took many years to develop to the degree it has today. MCI was not really on equivalent competitive terms with AT&T until equal access was implemented across the country. That process took a great deal of money, extensive network changes and software testing, and quite a bit of time to complete. Local competition cannot be expected to blossom overnight, everywhere, and in every market at the same pace.

The economics of the local market also do not yet encourage market entry in all market segments. Residential service remains subsidized, and in many areas the price currently set for the service does not encourage rapid entry. Pricing issues may become less of a problem once companies can offer packages of services and can build innovative offerings that include local service and long distance.

In fact, one very interesting way to encourage local entry might be to require that a flat-rate, stand-alone local service option be offered by all local competitors. The offering could be supported by universal service funds and perhaps targeted to customers with financial needs. It might even be set at a level that could not be changed or could go up only by the rate of inflation in a given year. Outside the one flat-rate option, however, companies would be free to offer any pricing combinations and packages of services they wanted. With the strong consumer demand for convenience, customer-tailored services, and innovation, competitors would see a market opportunity as a result of the proposal in many parts of the residential market that today look unattractive.

There is another reason for the impression that local markets are not opening fast enough—self-interest. Many of those who shout the loudest that local competition is not working are large companies that have lucrative long-distance businesses—worth over $70 billion domestically and billions more worldwide—they want to protect. Under the act, the Regional Bell Operating Companies (RBOCs) must open their networks to competition by satisfying a checklist of 14 items—everything from establishing local number portability to creating operating support systems that competitors can use to sign up new local customers of their own. They also must file with the FCC for an approval to enter the long-distance market; the FCC then considers comments from the states, the Department of Justice, and private parties in determining whether to authorize entry. The FCC is supposed to base its determination on whether the RBOCs have opened their networks to competition.

A Closer Look at Interconnection Pricing

A key issue in the equation is the price set for interconnection. Various forms of interconnection include unbundling, in which competitors use pieces of the incumbent's combined with their own networks to complete calls; resale, in which competitors simply resell the local network facilities of incumbents; and physical interconnection of the networks of competitors with those of incumbent carriers.

Again, Congress chose to rely on private negotiations between companies to ensure that proper and competitive prices between carriers developed. Negotiations can become mired on occasion so Congress also provided that where that occurred, the state commissions could act as arbiters. In such cases, the state commissions would set prices based on cost as required by the act.

In all the states in which it operates, Bell Atlantic began entering into interconnection agreements with competing carriers promptly following the passage of the act. Its actions helped establish an environment in which successful interconnection agreements were reached with scores of different companies, in different segments of the marketplace. Congress's belief that market-oriented incentives would set the stage for successful local interconnection negotiations is proving true in Bell Atlantic's region. In arbitrations and other state regulatory proceedings, the company has proposed prices that are based upon forward-looking economic costs. Under the terms of the merger commitments Bell Atlantic made to the FCC in July 1997, Bell Atlantic will continue that policy.

The reliance on private negotiations to promote efficient entry also reflects the strong desire of Congress to stimulate investment, innovation, and new network deployment. Interestingly, two other countries, both strongly committed to telecommunications competition in all markets, adopted policies designed to lead to investment in new infrastructure. Their approaches, while different, are achieving what Congress sought to achieve: a balance between the objective of encouraging early entry into local markets and the strong desire for investment in new network infrastructure.

In Canada, the government did not mandate local resale discounts. Competitors must pay retail prices for underlying services. Only essential facilities—those that cannot be economically provided in any other fashion—must be provided. Competitors in Canada do have access to unbundled facilities. but only a basic set of services must be unbundled. The government even moved to rebalance local rates, doubling them in some cases, without a significant drop-off in subscribership.

Canadian regulators recognized that new entrants come into a market with their own advantages and strengths, including smaller overheads, newer facilities and technologies, and often an ability to act more quickly to changing market conditions. Setting artificially low, regulated prices for interconnection is not necessary to stimulate entry, and will stifle investment.

The British, too, focused on encouraging facilities-based competition. They have not, for example, required unbundling of network elements and they have allowed cable companies and local telephone providers to package their services, increasing the incentives to invest in new services and infrastructure. The British system is certainly not perfect—British Telecom is not permitted to offer video services so markets in that country are not truly open to entry for all companies—but they have attempted to avoid much of the regulatory intervention that can skew market development.

The foreign developments have significance for the United States in the long term. To compete internationally, the nation must have a strong, innovative, and competitive telecommunications market and infrastructure.

In an effort to jump-start local competition, the FCC originally established so-called proxy rates for interconnection that were so low they would have discouraged competitors from investing. Rates set at those artificially low levels would have also discouraged incumbents from doing so. If every new service the incumbent companies develop can effectively be resold by competitors with little compensation in return, why would incumbents invest in such innovations in the first place? If incumbents must price access to their local networks at rates that do not reflect even their real forward-looking costs, including a reasonable profit, why invest just so competitors can use the fruits of investments at below-cost rates?

New broadband services that use existing local lines—such as the Digital Subscriber Line (DSL) family of services that provide high-speed data connections to the Internet—are now being readied for deployment by Bell Atlantic and other companies. If those arguing for below-cost unbundling and resale of services by incumbent telephone companies were successful in their efforts, the incentive to invest heavily in DSL deployment would be severely dampened. The creation of high-speed data networks would be slowed just as demand reaches high levels—and just as America's leadership in the Internet, the information network of the future, is reaching a peak.

The Long-Distance Picture

While trying to push regulators to adopt below-cost interconnection pricing standards, many competitors have at the same time argued that the states and the FCC should go slow in allowing the RBOCs into long-distance service. They claim that the RBOCs are failing to provide effective processes to interconnect networks and are working to stymie local competition. But industry analysts and experts do not agree. (Scott Cleland of Legg Mason, for example, commented recently regarding MCI's complaints about the RBOCs, "It's convenient to scapegoat both the Bells and the regulators rather than admit they're [MCI] in a rotten strategic pickle.")

The long-distance carriers have complained about the difficulty of local entry. Yet they have aggressively targeted the lucrative local business market—worth some $65 billion of the $105 billion total local telephone market—and high-volume residential callers, while virtually ignoring low-volume residential customers. Their commitment to the broader local market is suspect. In discussing their merger earlier this year, executives from Worldcom and MCI made it clear their strong focus would be on the business market.

Companies that are serious about entering local markets—Teleport and Metropolitan Fiber Systems among others—appear to be doing well, despite the complaints of the larger long-distance companies. In fact, Teleport saw its sales rise by 67 percent between the last quarter of 1996 and the first quarter of 1997. ACC, a New York-based competitive local exchange carrier, reported that its revenues from local and other services increased 58 percent in late 1997. Brooks Fiber reported a year-over-year local service revenue increase of 230 percent in 1997. A recent analyst's report projects that competitive local carriers will earn more than 5 percent local revenues nationwide this year.

Companies like those seem to be able to work with the incumbents as the new interconnection and operating systems begin operation. Responding to Sprint's claims that the incumbent companies were making it difficult to compete, the CEO of RCN, a new reseller, said, "Nynex's [information-sharing] systems are a little slow but they aren't so slow you can't do business." Bell Atlantic's local operating support systems are working, and the performance of the systems is improving every day. In fact, Bell Atlantic's progress in opening its network to competitors convinced New York regulators, with the concurrence of the U.S. Department of Justice, to approve a work plan intended to test Bell Atlantic's operating systems. The intent of the plan is to work through any problems and arrive at a point where the Commission can recommend to the FCC that Bell Atlantic should be allowed entry into long distance in New York. Even the Consumer Federation of America voiced support for the plan.

Toward Competition in All Sectors

As a country, we must encourage competition in all telecommunications markets, not just in the local segment. We must do so without artificially hamstringing either incumbents or new entrants to stimulate strong investment and market-based entry. That is especially important given the huge demand for bandwidth that has developed. How can we achieve those goals?

First, Do No Harm

Congress understood that relying on the private sector to the greatest extent possible would be the best way to ensure that viable markets based on sound economics and consumer demand would develop and expand. Some basic ground rules do need to be established, but intervening constantly in the market—especially in the setting of prices—is a recipe for continued regulation, not expanding competition.

In fact, economic regulation should diminish rapidly and disappear as markets open to competition. Regulation of private enterprise has a long history in this country and the unfortunate fact is that, when regulators are present in a market, companies of all kinds find it convenient—and often easier—to go to the regulators to determine market questions rather than to rely on market processes. Regulators will need to continue enforcing antitrust and competition laws and consumer protection policies. They should not continue to get in the middle of market entry and pricing questions.

Open All Markets to Entry As Rapidly As Possible

Congress wanted to encourage rapid growth in competition. The best way to do that is to open markets to entry rapidly so competitors can begin to operate. Opening the long-distance market to RBOC entry is the best means of convincing long-distance carriers to stop protecting their markets and begin aggressively competing in the local market. Given current consumer demands for convenience, value, and innovation, all competitors realize they must have a broader and varied portfolio of services to be successful.

Encourage Innovation by Allowing New Services to Be Offered Freely and without Regulation

Unlike countries such as Canada and Great Britain, the FCC has pushed unbundling to such a level that very small parts of the local network can be "rented" from the incumbent carriers. Many of those services can be programmed into switches, including those of competing companies, so there is no reason that the companies could not offer such services on their own.

Further, if local telephone companies are forced to offer immediately any new services they develop—such as high-speed, high-bandwidth services like DSL—on an unbundled basis to competitors, their incentive to invest in such innovations is greatly undermined. Because such services are new, consideration should be given to either deregulating or "walling them off" from unbundling for a period of time.

Where There Is Proof That Companies Are Intentionally Inhibiting Competition, Punish Them

Accusations and innuendo regarding anti-competitive actions in any market are not helpful and do not promote competition. Solid evidence that a company is intentionally inhibiting competition should be brought forward and dealt with by regulators and the antitrust laws where appropriate. Otherwise, bringing such accusations without evidence simply harms corporate image and reputation, damages consumer confidence in the entire industry, and undermines the integrity of the market.

5. Encouraging Infrastructure Investment: A Comparative Study

Henry Geller

Developments in the United Kingdom and Canada call into question the course of interconnection regulation taken by the United States and militate strongly for a modest course correction at this time. I will briefly review the background of recent telecommunications policy in the United States before turning to my main thesis.

The main purpose of the 1996 Telecommunications Act is to have telecommunications—a tremendous enabling technology—make a maximum contribution to the efficiencies needed in this era of global competition and to the quality of life in sectors like education, health care, energy conservation, and the democratic process. Telecommunications technology cannot make that contribution without moving in a timely fashion to advanced capabilities at the local level, where the information superhighway becomes a "dirt road."

Interconnection versus Infrastructure at the FCC

Thus, Section 254(b)(2) of the act sets out as a guiding principle that "access to advanced telecommunications . . . should be provided to all regions of the Nation." Section 706 requires the Federal Communications Commission and state commissions to encourage the timely and reasonable deployment of advanced telecommunications services to all Americans by using "methods that remove barriers to infrastructure development," including price caps, regulatory forbearance, and competition.

The August 8, 1996, Interconnection Report was by far the most important FCC undertaking to implement the 1996 act. The report followed Section 251(c) of the act in prescribing essential unbundled elements and the resale of incumbent local exchange carrier (ILEC)

Henry Geller is a communications fellow at the Markle Foundation and the former director of the Washington Center for Public Policy Research at Duke University.

telecommunications services at wholesale rates (retail prices minus the costs that will be avoided, estimated to be about 20 percent off the retail price). Indeed, the FCC pushed the regime to the limits and possibly beyond, in light of the Eighth Circuit Court's July 18, 1997, decision, now on review in the Supreme Court. The commission ordered the ILECs to satisfy competitors' requests for all unbundled network elements (UNE) without themselves supplying any facilities (so-called UNE platform or shadow networks), thus fostering arbitrage between such networks and the ILEC's retail pricing. To price the unbundled elements, the commission used a total element long-run incremental cost (TELRIC) pricing scheme, under which prices are based on the forward-looking costs of a hypothetically efficient network (roughly estimated to be 50 percent off the retail price).

The FCC would argue that the rules promote section 706. True, it has not used forbearance, price caps, and so forth, but it has focused on fostering competition. The commission's position, simply stated, is as follows: Competitors need access to the customer, and "resale" (in quotes to denote wholesale resale or a "UNE platform" resale) gives them that crucial access. Over time, they will build out their own modern networks as they do not wish to be dependent on the ILEC, and the ILEC will be forced to respond to this modernized competitor by in turn investing in advanced capabilities.

The ILECs counter that under this scheme, the FCC is discouraging facilities-based competition for the residential customer; most new entrants will simply use the ILECs' facilities, especially the local loop because it is available at such a cut-rate charge. Competition for the residential customer therefore will take place largely at the retail pricing level. Indeed, competing local exchange carriers (CLECs) such as the cable companies and Teleport cautioned that the FCC's proceeding in this fashion could lead to a *tsunami* of retail pricing competition that would thwart the CLECs' own facilities-based efforts. Further, the ILECs argued that not only would the local loop remain the dominant facility well into the next century (until effective wireless competition arrives); they will be discouraged from offering advanced capabilities. Why should they invest in advanced infrastructure for all Americans when it must be made available to competitors at the reduced "resale" charges?

The jury is still out as to whose position will prove to be correct. Newspaper and trade accounts deplore the absence of local competition. But the FCC is correct in saying that this is an evolving process and that it is unreasonable to expect strong local competitive developments within such a short time after the act's passage.

Interconnection in the United Kingdom and Canada

Significantly, two countries, the United Kingdom and Canada, both strongly committed to breaking the local monopoly's bottleneck, have decided to take a markedly different course than the United States. Both countries recognize the importance of effective interconnection policies to promote local competition. The question is what policies to adopt for the purpose. The director of Britain's Office of Telecommunications (OFTEL) recently stated that the United Kingdom did not adopt the U.S. scheme (in particular, unbundling) because the United Kingdom wanted to promote facilities-based competition rather than to opt for largely retail price "resale" competition. Ultimately, he explained, that would allow regulation of the local loop to come to a timely end. He further noted that facilities-based competition, especially by the cable operators, was making significant inroads in local competition.

Canada also has embarked upon a different course, as described in the June 23, 1997, issue of *Telecommunications Report*, page 14:

> First, it didn't mandate local resale discounts; competitors must pay retail prices for underlying services. Second, in Canada, only "essential facilities"—those that can't be provided economically in any other fashion—must be provided. In practical terms, only three network elements must be unbundled: access to telephone numbers, access to directory listings, and local loops in high-cost areas. In a concession to facilitate early market entry, however, the Canadian Radio-Television and Telecommunications Commission (CRTC) decided to mandate unbundling of all local loops during the first five years. Prices for unbundled elements will be based on long-run incremental costs, plus 25 percent for joint and common costs. Competitors in Canada weren't given access to incumbents' back-office operation support systems. . . . In decisions over the past four to five years, Canadian regulators have rebalanced rates, doubling local exchange rates in some cases. . . . But there has been no significant dropping off of subscribership on the public switched network.

These foreign developments have significance for U.S. policy in both the long and the short term.

Lessons for the Long Term

In the long term, we will have a chance to observe the results of competition among national telecommunications policies. There are, of course, differences among those nations that would affect the outcome of such competition. But if, at the end of a three-year period, the trend toward U.S. facilities-based competition for residential customers is dismal or inadequate, and the trends in the United Kingdom and Canada are promising in that respect, surely that militates strongly for a sunset of the U.S. scheme. That does not mean there would be no U.S. directives for interconnection and resale. Section 251(a)(b), requiring ordinary interconnection and resale, would remain and is applicable to all local exchange carriers. Rather, it would be the end of an extraordinarily detailed regulatory plan that had not proven out.

In three years, the FCC would have the power to act to remove that regulatory scheme under the forbearance provisions of Section 401. The commission need not forbear from price cap regulation (Section 401(a)(l)(2)), but only from applying the unbundling/wholesale resale plan to ILECs. This it could do under the public interest standard of 401(a)(3) and be consistent with 401(c), since by that time the requirements of the checklist would have been fully implemented (and found to be a failure so far as providing substantial facilities-based competition is concerned). Significantly, that action would parallel roughly the five-year period during which Canada mandated an unbundled local loop. Indeed, a strong argument can be made that the present complex interconnection regulatory scheme, successful or not, should be ended after an appropriate period (e.g., five years).

For the Short Term, Return to Section 706

As to the short term, the actions of two neutral and committed regulators like the United Kingdom and Canada raise doubt as to the efficacy of the U.S. approach and suggest that the FCC would be unwise to rely on interconnection alone to achieve the vital goal of Section 706. The commission should immediately remove a substantial barrier to infrastructure investment by making the unbundling / wholesale resale requirements applicable only to the existing network, not to future advanced capabilities. Indeed, Section 706 requires the FCC to take into account whether its interconnection report has imposed a substantial barrier to the timely deployment of advanced telecommunications capabilities and, if so, to remove it. The commission's interconnection report gives the ILECs a disincentive to invest in advanced telecommunications facilities, which must be made available to competitors under the unbundling / wholesale resale scheme, and an incentive to invest in deregulated areas, here and abroad. The requirements of Section 706 must be read in conjunction with those of Section 251(c), the unbundling / wholesale resale scheme. (If some element of an integrated advanced telecommunications network replaces an essential element of the present narrowband system, that element should continue to be made available under Section 251 but solely on the existing [narrowband] basis.)

I believe that Section 706 is controlling. The FCC should reach the same result if the matter were considered under the forbearance provisions of Section 401. Section 401(c) states that the FCC may not forbear from applying the requirements of Section 251(c) or 271 until it determines that those requirements have been fully implemented. Clearly, the act permits future forbearance when enforcement is unnecessary to ensure just rates, consumer protection, or consistency with the public interest (i.e., when some sector is in the effective competitive zone).

That is the case with future ILEC advanced telecommunications capabilities for residences. The ILEC has no present monopoly or market power in that area and indeed starts behind cable in video distribution or high-speed Internet connection, where it will face competition from several competitors, especially cable. An ILEC's separate subsidiary providing such broadband services should be completely deregulated. Again, Section 706's call for the use of forbearance to accelerate deployment of advanced telecommunications capabilities makes a most compelling case for this short-term reform.

This is a win-win situation to encourage infrastructure development. The ILEC no longer will be deterred from developing new vertical service or advanced capabilities like Asymmetric Digital Subscriber Line (ADSL) or hybrid fiber / coaxial (HFC) systems, because it must make them available to rivals under the 251(c) plan.

The CLECs, in turn, will have a strong incentive to develop advanced capabilities to meet or trump any such ILEC efforts (e.g., by developing their own new vertical services or adding ADSL electronics to the "dry copper" of the local loop). In either event, the public will be well served.

There could also be a gradual phaseout of the UNE/TELRIC regime as a reasonable way to provide incentive to both the CLEC and the ILEC to invest in infrastructure. The latter would have the assurance that the "UNE platform" would end at some fixed time, and the CLEC would have to develop plans and eventually build out its own facilities. In that respect, there could be a distinction between the local loop and other elements like switches, which are readily available from many competitive sources. After a reasonable time, why should not large carriers like AT&T and MCI be required to obtain their own switches?

That development would make a substantial contribution to facilities-based competition, because with the switch there is the possibility of innovation in vertical services, advanced intelligent networking, and so forth. Alternatively, the FCC could consider simply raising the price of all elements, after an appropriate period, on a gradual annual basis, until they reach a price level allowing ILEC recovery of historic costs.

In sum, the FCC should as expeditiously as possible adopt new policies to accelerate the deployment of infrastructure that will bring advanced telecom capabilities, so needed for efficiencies and to enhance the quality of life, to all Americans.

6. Mandatory Interconnection: The Leap of Faith

Solveig Singleton

Many observers of the telecommunications industry anticipate problems with the interconnection regime erected under the Telecommunications Act of 1996. Most critics focus on the Federal Communications Commission's choice of pricing models. But the problems with the mandatory interconnection regime are more fundamental.

First, neither Congress nor the policy community as a whole made clear exactly what the goals of the act's interconnection provisions are. Is the economic problem to be solved the new entrant's need for facilities or for access to the customer base? Was the regime intended to be phased out, or must it remain in place forever?

Second, Congress and the FCC adopted mandatory interconnection as a cure-all for sick telecommunications markets, giving little thought to the regime's drawbacks. While some of the drawbacks can be minimized, others resist rational solution. The situation suggests that the best long-term solution to the problem of interconnection is to phase out regulation.

Do New Entrants Need Facilities, or Customers?

The trouble with interconnection began when ordinarily ardent supporters of the free market fell all over themselves embracing the premise that competition would not arrive unless regulators required established local telephone companies to interconnect with new entrants. Electrified by the possibility of introducing competition into the local exchange, many commentators glossed over some important issues.

Solveig Singleton, a telecommunications lawyer, is director of information studies at the Cato Institute.

The argument for mandatory interconnection runs as follows: New entrants need mandatory interconnection and there will be no "competition," as the word is commonly understood, otherwise. Incumbent telephone companies in the local exchange have no incentive to bargain with new entrants for access to their networks and customers, because of the incumbents' large market share. E-mail systems interconnected voluntarily, and so did some competing local telephone companies at the beginning of the century. But those negotiations took place under two conditions: each carrier party to the interconnection agreement had some customers to begin with, and customers themselves did not have the settled expectation that all other customers would be on the same network. And, the argument concludes, the government is part of the problem, so it should be part of the solution. That, in a nutshell, is the argument for mandatory interconnection.

The argument betrays a disturbing fuzziness. Will new entrants need mandatory interconnection because they cannot afford to build competing *facilities*, or because even if they have built competing facilities, they have no *established customer base*?

The rise of MCI in long distance and the competitive access providers in the local exchange shows that the problem of building competing facilities is not an insurmountable one. With cable networks already in place (albeit requiring conversion to two-way) and wireless facilities growing, access to facilities is not the fundamental economic problem facing new entrants.

The real problem for new entrants is that even with a fully constructed two-way network in place, a new entrant would have a hard time getting new customers except on a limited private-line basis, because no one will want to subscribe to a network that cannot be used to call somebody on another network. Solving this economic problem, which I will call *access to customers,* will sometimes entail access to some "essential" facilities. But the two economic problems nonetheless are distinct.

So the debate over interconnection in telephone markets is different from and more difficult than the debate over interconnection in railways or electricity. If one accepts this argument, it follows that our current regime of mandatory interconnection, which seems designed to provide access to facilities even when that is not necessary to give access to customers, is broader and more intrusive than it needs to be. Because it is being directed at solving problems that do not necessarily need to be solved, it may miss solving the problem that does need to be solved.

The arguments for mandatory interconnection described above conceal other ambiguities in the goals of interconnection regulation. Telecommunications scholars would agree that "competition" is good. But they would differ widely as to what "competition" in telephone markets ultimately should look like.

Many commentators take as their ideal a state of "perfect competition," where numerous firms of roughly equal size compete happily in endless equilibrium, while customers enjoy perfectly cost-free information. While grudgingly admitting that real-world markets lack those characteristics (and never asking whether it makes sense to choose a world so unlike our own as a normative standard), many of the commentators at least would hope to see more than one entrant in each local telephone market. Those who foster that hope may be assigning the interconnection regime a wholly unrealistic goal that bears little relation to free markets.

The main obstacle to achieving the goal is the "network externality," that is, the fact that the network increases in value by adding more customers (by interconnection, buildout, or merger between companies)—that is, by becoming *big*. Perhaps even if the government had never been involved in telephone markets, markets today might be relatively concentrated.*

Now, many would take that to mean that interconnection regulation is indeed necessary and, what is more, that it must be permanent. It is not merely a transitional measure.

But note that if it is indeed unlikely that in a truly free market the phone market would not much resemble markets for, say, shoes or groceries, *any regulatory solution designed to produce that effect will be fighting a losing battle*. Whatever benefits the regime generates will be benefits not of a free market but of managed competition. There is no point in pretending otherwise.

We have a choice at this point. We can commit ourselves to fight that losing battle. Or we can face our deep-seated horror of monopoly

*In conversation, Bob Pepper of the FCC noted that if markets were starting from scratch with today's wireless technology, they almost certainly would be very competitive. I agree but note again that in the real world one never starts from scratch.

(or even worse, *natural* monopoly), head on, and admit that we know less about the problem than we think we do. Could market power in the real world turn out to be a less fearsome thing than regulatory power—at least from the standpoint of consumers, if not competitors? The markets dominated by Microsoft and Intel, respectively, by almost every measure serve customers well.

On the other hand, we have the standard of "perfect competition," compared to which monopoly is inefficient. These models of what competition should look like began as academic exercises with some predictive or explanatory power. But those static models were not intended to serve as a normative standard by which to judge developments in the real world, and they do not seem to apply at all well to fast-changing telecommunications markets.

So we may discover that there is an alternative to fighting the losing battle of regulation—at least, in the long run. We really do not know, but have only assumed, that a world without mandatory interconnection would be a terrible place. In my view, there is a better alternative than endlessly fighting the losing battle of regulation, a world without any mandatory interconnection in it at all.

What Are the Drawbacks of Mandatory Interconnection?

Mandatory interconnection was adopted as a solution to the problem of the local exchange without anyone seriously considering what the drawbacks of mandatory interconnection are.

To start, there are two familiar objections to our current interconnection regime. The first is that the FCC picked the wrong pricing model, thus stunting the future growth of networks. The second, related problem is that the rules violate the takings clause of the Constitution. The response to both is the same: the FCC must set the right price. The reality, though, is that the FCC will never and could never get the pricing right. Historic cost is not the right pricing model (although one can argue, as Professor Richard Epstein has, that only this model avoids a takings problem). The right model is market pricing. But market pricing in the real world includes factors such as market power—and the entire purpose of the interconnection regime is to defeat market power.

And while market pricing in the real world tends to follow costs, that does not mean that costs are the basis of market pricing. Costs and prices have a chicken-and-egg relationship. Businesses look at what their competitors are charging; they look at what the market will bear; if they do look at their costs, it is partly with an eye to recovering them and partly to see if they can reduce them or otherwise change them in any way. Costs change day by day, region by region, block by block.

So the whole attempt to model pricing on costs is flawed. That means that mandatory interconnection will always distort the growth of networks and raise significant takings clause issues. The regulators cannot get the prices right.

Next, mandatory interconnection entails continued regulatory involvement. What is more, it creates special interests that favor continued regulatory involvement. The mandatory interconnection and resale regime for long distance created hundreds of tiny resellers, entirely dependent on the FCC and endangered by any movement toward further deregulation. Even if policymakers decide that interconnection should be a transition to free markets, we may suppose that it will be a permanent feature of the legal landscape.

Finally, mandatory interconnection is premised on the view that it will be reasonably efficient in practice to force unwilling business partners to deal with one another. This vastly underestimates human potential to cause problems. Accusations already abound. Ray Smith of Bell Atlantic suggests that long-distance companies are holding back from entry into local telephone markets to keep the local companies from offering long-distance service. Long-distance companies and new entrants to the local exchange accuse local exchange carriers (LECs) of lengthy delays in processing interconnection orders. One attendee at the conference suggested that some local exchange companies were attempting to design a switch that could not be unbundled. Whether one believes some or all of the accounts, all are plausible. Can one really erect the telecommunications infrastructure of the future on a foundation of ill will?

Some Solutions

What new entrants need is access to customers, not access to new facilities. Recognizing this gives us an opportunity to review every item that the FCC ordered unbundled and resold, and ask whether it is really necessary to solve the problem of access to customers. If not, it should not be commandeered by regulators.

More particularly, we should take a closer look at mandatory resale and prepare to scale it back or eliminate it entirely. Years ago in a proceeding that involved cellular entrants' rights to resell incumbents' services, the FCC noted that resale was intended to give the new entrants time to build up their own networks. But it was not working that way. Instead, the new entrants simply positioned themselves as perpetual resellers. Steps should be taken now to avoid that problem in the local exchange.

Next, how can we stop the interconnection regime from creating a class of new "competitors" entirely dependent on the FCC? One way to make it clear that new entrants must take steps to develop their own networks is to impose a sunset date on the entire interconnection regime. The sunset provision could be effective either on a certain date or (to avoid the problem of incumbents' ability to impose deliberate delays that David Turetsky of Teligent pointed out) when local incumbents have lost a certain percentage of market share.

In addition, we can recognize that regulators are not going to get the pricing "right," so perhaps they should no longer try. Instead, the interconnection regime should be redesigned to ensure rough-and-ready interconnection agreements that permit some kind of competition at any level as quickly as possible. That means a regulatory regime that relies much more heavily on arbitration, rather like the approach that Pablo Spiller and his colleagues designed for Guatemala. Arbitration also might help to end strategic delays.

Toward Free Markets

The next step in the quest for free telecommunications markets is to understand exactly what economic problem new entrants face in the local exchange and which rules move the industry toward a long-term solution to that problem. It is possible to design a mandatory interconnection that is less interventionist—and more effective—than the current system. In the long run, however, mandatory interconnection should be done away with entirely. Only then may we welcome the reality of a free telecommunications market, without expectations distorted by unreal economics models and without markets distorted by political intervention.

7. The Unbundled Road to Regulatory Reform

Peter K. Pitsch

The interconnection issues raised by the Telecommunications Act of 1996 are among the most challenging and important policy and legal issues that the Federal Communications Commission and state regulators have ever had to face. First, the underlying economic questions are formidable. Second, the act is ambiguous about the scope of the FCC's jurisdictional authority and which methodology should be used to estimate costs. Third, many of the competing considerations are difficult to balance. For example, implementing the act entails weighing the benefits from decentralized decisionmaking by state regulators against the efficiency and competitive gains to be had from setting uniform national guidelines. Finally, the dynamism of the telecommunications industry and linkages to key issues such as reforming local access charges paid by long-distance carriers and making universal service subsidies competitively neutral complicate resolution of interconnection issues. Not surprisingly, the pleadings filed with the FCC and the Eighth Circuit Court of Appeals on an array of interconnection questions are enough to boggle the minds of even sophisticated observers.

Although state utility commissions and courts considering important elements of the FCC's August 1996 decision have made some headway, the interconnection debate is still a work in progress. Even if the Eighth Circuit's decision on jurisdictional issues is accepted as the final word (which may not be the case), state regulators must wrestle with the same difficult pricing issues. They are free to make, and many of them are making, decisions similar to those made by the FCC. But those state proceedings have a long way to go. While

Peter Pitsch, chief of staff to the chairman of the FCC from 1987 to 1989 and chief of the FCC's Office of Plans and Policy from 1981 to 1987, is an adjunct fellow at the Hudson Institute and a telecommunications attorney.

more than 20 states have adopted a forward-looking cost methodology for pricing so-called unbundled network elements (UNE), what precisely that will mean is not yet clear. What model will they use to measure the forward-looking costs for UNE? More important, how will the model be applied? What portion of the costs of providing UNE will be determined to be common and how will the costs be recovered? Will any of the local telephone company's costs be judged to have been stranded by competition, and how will those costs be recovered? How extensively will the states require the local telephone companies to unbundle their networks? At what level will the state regulators set the wholesale discount, and so on.

Given the difficulty and extent of their tasks, what interconnection policies should regulators adopt? I will develop two fundamental points in this essay. First, I will explain why the proper implementation of UNE represents the most practical and potent means for reforming telecommunications regulation. Second, I will describe in broad brush strokes what a proper UNE approach would look like.

The Public Choice Perspective on Regulatory Reform

Understandably, resolving the interconnection issues creates considerable unease for anyone with strong free-market predilections. It seems at odds with regulators' abandonment of cost-plus regulation and the adoption of incentive-based modes of regulating local telephone companies. Setting prices based on costs is a complicated regulatory process. It is also an intrusive process that if done improperly could undermine the incumbent telephone companies' incentive to invest and innovate. Nonetheless, I believe an enlightened implementation of UNE offers the best available road to regulatory reform in the local telephone market.

During my years in various positions at the FCC I found that the economic interests and the policymakers involved usually understood well the competitive impact of regulation. I also found that in the end, talking about deregulation did not in itself generate much deregulation. More important were certain conditions and factors that fostered deregulation. New entry into previously regulated communications markets was the single most important driver of the deregulatory process.

The long-distance market is a three-decade-long example of this process. In the initial phase, the FCC allowed entry. For example, in 1959 businesses were permitted to build their own private microwave networks (Above 890) and in 1976 MCI was permitted to provide regular long-distance service (Execunet). That entry inexorably forced regulators to allow incumbents to "restructure" their rates, that is, adopt more efficient prices. Telpak, Reach Out America, and True USA provide some of the more notable examples of competitive responses. Eventually, competitive entry led to the substantial deregulation of long-distance carriers, including AT&T.

The deregulatory process occurs in this fashion because the merits of competition and consumer choice are an easier political sell than the merits of efficient prices and deregulation. (This makes sense given the everyday experience of benefiting from having a choice versus the much rarer pleasure of learning about the merits of efficient pricing and reductions in the often-hidden costs of regulation.) Also, deregulation often raises reasonable concerns about the exercise of monopoly power on the part of an incumbent company that is deregulated before competition develops.

Whatever its rationale, the entry/rate restructuring/deregulation process is about to be replayed in local telephone markets. The Telecommunications Act of 1996 is far from perfect. It failed to address the inequities and consumer welfare losses due to current pricing distortions of local telephone service. (Crandall and Waverman estimate that correcting the distortions could generate an annual consumer welfare gain of $8 billion.)[1] In fact, in several cases it took us back a few steps. I have in mind the codification and extension of subsidies to favored areas and groups. The saving grace of the Telecommunications Act, however, is its emphasis on entry. As in the past, entry will be the real driver of efficient pricing and deregulation.

From that perspective, I see interconnection regulation, in particular the proper implementation of UNE, as the best available approach to regulatory reform in the local telephone market. Congress, the FCC, and the state utility commissions were unlikely to initiate substantial rate reform on their own. By promoting entry, however, the effective implementation of the UNE option will provide the "necessity" to "mother" reform. It will

- foster efficient facilities-based entry,
- promote constructive solutions negotiated between incumbents and new entrants,

- expedite adoption of efficient intrastate and interstate access rates,
- promote the competitively neutral recovery of universal service and other subsidies, and
- foster deregulation of retail telephone rates.

The Unbundled Road to Regulatory Reform

To best achieve the competitive, rate-restructuring, and deregulatory benefits of new entry, regulators need to adopt the proper UNE system. In general, they should strive to adopt a system that efficiently prices unbundled access to the elements of the existing network. Unlike the other resale options available under the Telecommunications Act of 1996, UNE prices are more likely to send the proper signals to new entrants about the relative efficiency of using existing facilities versus building their own. Experience in the long-distance market has shown that permitting new entrants to resell the incumbent's network can be an effective means of reducing their entry costs. In this way entrants can add facilities in an incremental fashion.

Access to the incumbent carrier's network is likely to be even more important in local telephone markets, given the sheer size of the necessary expenditures, the increasing doubts about the extent to which facilities-based competition is currently feasible, and the potential importance of providing customers local and long-distance services in a convenient one-stop package. UNE regulation is the most practical solution. At the same time, UNE regulation should permit incumbent local telephone companies to restructure their rates and to benefit from cost-reducing and innovative efforts. The following guidelines would promote the proper implementation of UNE regulation.

UNE Prices Should Be Based on Forward-Looking Costs

Forward-looking prices for UNE will promote efficient entry and investment. Investment decisions should not be based on historical or accounting or past regulatory costs, because that would distort the prices of final goods and act as a barrier to efficient entry to the detriment of competition and consumers.

For competitive neutrality reasons, I do not favor the use of auctions to set universal service subsidy levels. Yet the proposed use of auctions to set such subsidies illustrates the logic of basing UNE prices on forward-looking costs. Actively contested auctions would set the universal service subsidy at a "competitive" level—one based on forward-looking estimates of costs. The historical or accounting costs of an incumbent local telephone company would be largely irrelevant in such an auction. By necessity, the incumbent local telephone carrier would compare its forward-looking costs with those of its likely competitors. If its forward-looking costs were lower than its competitors', then it would be better off winning the auction even if to do so it had to forgo recovery of some of its historical or accounting costs.

UNE Prices Should Not Be Based Solely on "Green Field" Estimates of What an Efficient Network Would Look Like

The process of setting forward-looking prices for UNE should recognize the limits of the regulatory process, especially those of the numerous technical and business considerations that go into creating a working network. If regulators consider the forward-looking costs of purely hypothetical networks, they may make costly mistakes. Such mistakes could undercut the incentives of competitive local exchange carriers to add innovative facilities where that would be efficient. In this regard, I think the FCC's interconnection order got it right. Regulators should base UNE prices on forward-looking costs for the incumbent local telephone company's network, assuming current wire center locations.

Recovery of Common and Stranded Costs Should Be Competitively Neutral

Forward-looking pricing for UNE does raise issues of how to recover "common" and "stranded" costs. Common costs are those overhead costs necessary to the provision of multiple unbundled network elements. Stranded costs are those material costs, if any, that the incumbent carrier had to incur to meet a regulatory-approved end and that it has been unable to recover in the newly competitive market.

One advantage of UNE regulation is that common costs should be much smaller than they are for the price regulation of individual services, because many of the costs can be ascribed to individual unbundled network elements. In contrast, the different kinds of services typically use the entire network and therefore there is no basis for ascribing the common costs of the network to individual services. Whatever costs regulators find to be common to the provision of UNE should be recovered in a competitively neutral manner. Placing them directly on the subscriber in the form of an end-user charge would probably be the most efficient means of recovering them. Any subscriber of the incumbent carrier or a competitive carrier using unbundled network elements would have to pay the charge. Customers of a facilities-based competitive carrier, on the other hand, should not have to pay the charge, and it could be adjusted to the extent that a carrier used its own facilities.

Regulators should also allow carriers to recover any stranded costs in a competitively neutral manner. State regulators should determine to what extent the incumbent carriers have incurred material costs to meet regulatory goals or requirements and have been unable to recover those costs in the new environment. The process needs to be open and accountable, and the incumbent companies should bear the burden of proof. Again, the best means of recovering the costs would be through an end-user charge on all subscribers. In this case subscribers of all carriers including facilities-based competitive carriers would pay the charge. (The same should be true for the funding of universal service subsidies.)

Regulators Should Foster Rate Restructuring

Regulators should encourage the efficient restructuring of local rates by allowing incumbent carriers to match prices adopted by new entrants. That is, they should be able to match any new carrier's price cut as long as they can reasonably demonstrate that the matching price cut would be profitable. Where the new entrant is relying primarily on UNE, the matching price of an incumbent carrier could be presumed to be profitable.

On the other hand, new entrants should be able to use UNE to provide local access to long-distance companies. The reform is essential to eliminate inefficient subsidies of local telephone rates by long-distance callers.

Marketplace Incentives Should Be Strong

Regulators should implement UNE to minimize regulatory interference with the normal marketplace incentives for incumbent and new competitors. Regulators should allow parties to negotiate interconnection deals on a customer-specific basis. The presence of a regulated UNE option should provide the needed offset to the incumbent carrier's monopoly position. Because new entrants will consider UNE a reasonably close substitute for wholesale rates based on so-called top-down estimates of retail costs avoided, the pricing of UNE on forward-looking costs will create a constructive environment within which negotiations can take place. Thus, with a boundary-setting guideline on UNE, regulators should quickly approve interconnection deals negotiated between incumbent carriers and new entrants.

Having determined the appropriate initial UNE prices, regulators could impose a price cap formula. They could revisit the formula in four to five years to determine if UNE regulation remained necessary. UNE regulation should be phased out when effective facilities-based competition exists. During the review, if regulators found that UNE regulation remained necessary, they could determine then whether the formula needed revision. In that way the advantages of incentive regulation could be combined with the benefits of the regulation of unbundled network elements. For example, under a price-cap approach the incumbent carrier would have normal business incentives to innovate and cut costs.

Last, regulators should adopt a UNE proposal that provides the basis for deregulating local retail rates. For example, incumbent local companies should be deregulated when competitive facilities-based and UNE-based competitors meet specified availability (or "passby") and penetration tests. Experience with cable regulation and different measures of effective competition has demonstrated the need to consider competition at the submarket level. However, a passby/penetration test would provide incumbent carriers with an added incentive to implement UNE regulation successfully as well as eliminate retail regulation when it was no longer necessary.

Conclusion

The plight of the deregulator all too often seems like that of the lost visitor who upon asking a local inhabitant for directions is told, "You can't get there from here." In telecommunications and other sectors, the road to deregulation has been proven time and again to start with new entry. The journey in local telephone markets will not be a simple one. Yet, given the circumstances of those markets and current political constraints, I believe that the surest and best road to regulatory reform will be an unbundled one.

Note

1. Robert W. Crandall and Leonard Waverman, *Talk Is Cheap* (Washington: Brookings Institution, 1995), p. 94.

Part III

Speculating in Spectrum: The Future of Property Rights

8. Underregulation: The Case of Radio Spectrum

Thomas W. Hazlett

What everybody believes is never true.

Friedrich W. Nietzsche

The 1996 Telecommunications Act: Speechless on Wireless

The "comprehensive" and "sweeping" Telecommunications Act of 1996 left one-half of our telecom marketplace untouched by legislative reform. Great commotion accompanied the rules enacted by Congress for the regulation of the *wireline* communications systems, particularly with regard to the interconnection of local exchange and long-distance telephone companies. Meanwhile, this once-in-a-lifetime opportunity to revamp the rules governing our *wireless* technologies—a regulatory system concocted in 1927—slipped past with scarcely a yawn.

The one visible margin on which the issue of spectrum regulation reform arose was within the context of the broadcast license "giveaway," as then-Senate Majority Leader Robert Dole called it. Since the Omnibus Budget Reconciliation Act of 1993, federal law in the United States has singled out television and radio stations for special treatment: only *broadcasters* will be awarded new wireless operating licenses by the Federal Communications Commission without being subjected to competitive bidding. The 1996 Telecommunications Act actually *strengthened* the broadcast license exemption, while reform efforts fizzled.

In the consensus opinion of those economists who have studied it, the U.S. spectrum allocation system does not well serve the interests of consumers, competition, or innovative technological development. Indeed, the system has, from its earliest roots, settled on the

Thomas W. Hazlett is a professor in and the director of the Program on Telecommunications Policy, Institute of Governmental Affairs, University of California, Davis.

protection of producers at the expense of customers, of regulated incumbents to the exclusion of unregulated competitors, and of existing know-how at the expense of imaginative new communications systems.

While reforms have, since the mid-1970s, slowly loosened many of the rules within this rigid system of wireless licensing, the central structure remains in place. That structure continues to impede rivalrous market forces, imposing large costs on the efficiency of the telecommunications sector in the United States. The source of this inefficiency is the fundamental top-down structure of the regulatory system: agency personnel are legally obligated to allocate radio spectrum via administrative process. The standard for decisionmaking (born in the 1927 Radio Act) is "public interest, convenience, and necessity," which is no standard at all. Every action by an administrative agency is—by definition—undertaken according to "public interest, convenience, or necessity." The requirement that agency personnel choose among countless possible services, technologies, and market structures suffers from the inherent limitations of socialism. The information relevant to discovering the optimal allocation of radio spectrum is nowhere available to the regulator, and preempting (in favor of "public interest" allocation) the competitive profit-seeking race to deploy spectrum in its highest and best use banishes the data necessary to solve the problem.

The process by which the FCC decides what spectrum will be used for what service, and how that service will be delivered, is necessarily dependent on the information held by private parties. Those best able, not to say most eager, to come forward to reveal such information to the authorities are not a random sample of the entire range of telecommunications providers. The sample, and hence the information provided, is overwhelmingly biased in the direction of regulated industry incumbents, firms that have a sharply defined interest in actively contesting any allocations that disturb existing profit streams. This situation creates an imposing burden for aggressive entrepreneurs, new technologies, and pro-competition regulators. Even a step in the direction of pro-consumer reform is bound to raise a flurry of administrative process, raising the cost of such reforms and discouraging agency personnel from heading in such directions to begin with.

Pro-consumer reform would require elimination of the procedural barriers that block access to radio spectrum by new entrants or that limit innovative uses of frequencies already assigned. Rather than dictating spectrum use in a top-down framework where whatever is not permitted is forbidden, a liberal regime would permit whatever is not forbidden—where what is forbidden is interference with existing transmissions. Services, technologies, and standards would not be set via administrative processes at the FCC, although industry groups would be permitted (shielded from antitrust liability) to craft voluntary standards. Such a system would encourage the deployment of any wireless service that could compete successfully for radio waves in the marketplace. Spectrum then indeed would be allocated to its highest valued use. Moreover, the artificial scarcity imposed via regulatory constraints would be lifted; spectrum would be cheaper—and quicker—to access. That would encourage waves of wireless research, experimentation, and deployment throughout the communications marketplace.

The FCC License Auction Reform

In 1993 Congress overcame its long-held opposition to competitive bidding mechanisms and authorized auctions to assign nonbroadcast licenses. But the initiation of auctions, and the approximately $20 billion in revenues they brought to federal coffers in the 1994–96 period, did not fix the spectrum allocation system. Indeed, FCC licenses are auctioned in the marketplace only after the agency has administratively determined what service is to be provided in a given band. Incumbents continue to hold up access to the airwaves by dominating the administrative process, denying consumers competitive services and innovative products.

The auction concept should be extended to the spectrum allocation process. Congressional reforms are needed to allow competitive entrants the right to access radio spectrum not currently in use. Where competitive claims are made, bidding mechanisms can quickly and profitably (in terms of government revenue generation) select claimants.

The key regulatory focus of such a regime is radio interference. Arbitrating technical disputes can be a smooth process, or it can be a nightmare—depending on the gains the participants anticipate from strategic behavior. When applicants show up at the FCC offering an exciting new technology, the reflexive response of threatened incumbents is to raise the specter of "technical reasons." Under the "public interest" rulemaking process, claims can be made at little cost and the burden of proof is on the entrant. Meanwhile, incumbents benefit from each and every delay. In this environment, many questions are raised, many answers disputed, and many a competitor crushed.

Under tighter rules with symmetric accountability, interference issues can be hammered out in straightforward ways. Indeed, the FCC dates its efforts to liberalize spectrum use to the mid-1970s when it opened up the Common Carrier Point-to-Point Microwave (CCPM) band. Prospective users of the band were presumptively allowed access to the radio waves, provided only that they not interfere with existing users. The regime set virtually no rules about how that would work. In a 1986 staff paper from the Office of Plans and Policy, an FCC engineer explained this remarkably liberal regime:

> Perhaps most notable of all is the absence of even a working definition of harmful interference. . . . [W]hile the Commission has not required it, a consensus appears to have emerged for adherence to a single, uniform set of interference criteria as a voluntary standard in order to facilitate the general coordination process.

When regulators are limited to arbitrating interference disputes, in a role more judicial than administrative, tremendous competitive forces will be unleashed to the obvious advantage of American consumers and businesses alike. License auctions do not, themselves, produce such fundamental reforms. License auctions merely capture the estimated present value of future profits for the federal government. They do not reveal the social value of radio spectrum or allocate frequencies to their highest valued uses. The term "spectrum auctions" thus misleads. The FCC merely auctions a license to operate within the rules established by its prior allocation of spectrum.

Spectrum reform should convert the current FCC license from an operating permit into a spectrum license. The CCPM liberalization discussed above made some progress in that direction. The broadband Personal Communications Services (PCS) allocation continued the trend in that it defined service broadly, did not establish a transmission standard, and auctioned regional licenses simultaneously to allow geographic aggregations according to market dictates. Most important, perhaps, the PCS allocation implicitly instituted a regime of *voluntary reallocation* of radio spectrum. Instead of clearing the PCS band of incumbent microwave users, the FCC issued licenses with "overlay rights." PCS providers had the right to use allocated bands so long as they refrained from degrading the microwave transmissions of existing (non-PCS) users. Some occupants elected to vacate the PCS band and relocate to higher frequencies as a result of bilateral negotiations (in essence, payments to make spectrum available to the PCS operator), while incumbents faced incentives to leave particularly valuable (costly to move) users in place.

As spectrum regulation is brought in line with the standard body of law, including rules of liability, contract, property, and antitrust, spectrum resources will contribute more and more to economic growth. This normalization process, by separating the electronic media from the current collusive chicaneries of press licensing, will also increase the health of our democratic institutions.

Calls for Spectrum Scarcity

Recently, policymakers and lobbyists in the telecommunications debate have argued that the government should manage the spectrum so as to keep demand for licenses—and government auction receipts—high. In fact, that is upside-down public policy. Because the radio spectrum is a nondepletable natural resource, we can only waste it by not using it. The government's restriction of radio spectrum may increase auction revenues, but it leaves American consumers and businesses facing higher prices and poorer choices in wireless services. Indeed, it is not simply a question of transferring our profits (auction receipts) from one pocket, to cover the losses (higher prices due to less competition) in the other. The efficiency losses to the U.S. economy can be staggering, particularly with respect to the new wireless services that such restrictions suppress.

There are three ways to withhold radio spectrum effectively. One is to allow a given band to lie fallow in whole or in part. A second is to allocate a given band to an entity that has weak incentives to make effective use of the valuable resource for the public. That happens when, for instance, the Department of Defense is given overly generous allotments. A third restrictive mechanism is for allocation rules to impose such inflexibilities that licensees cannot most effectively meet consumer demands, thereby locking in existing services and obsolete delivery systems.

Rationalizing our use of the radio spectrum resource is thus a three-pronged war, the goal being to expand wireless communications opportunities for the public. Promoting greater, and less constrained, access to the frequencies will maximize the *social* value of radio waves, where social value includes both producer and consumer gains. That optimum will, in fact, tend to reduce producer gains toward zero—a happy outcome in that spectrum is an input, not an output. When the cost of an input falls, economic welfare rises.

Spectrum's Value at the Margin

Wires used for communications constitute "spectrum in a tube." Wireline and wireless are technical substitutes—and have been for over a century. At its birth, radio was christened as a competitor to the telegraph. The fledgling nature of the cellular/landline rivalry today only hints at the possible efficiency gains that liberalizing spectrum policy could deliver.

By way of example, hypothesize full use of the so-called 402 MHz TV band—most of which delivers neither a TV signal nor any other telecommunications service to the public. Given the very impressive growth in demand for Internet service, and the growing frustration of computer users facing slow access speeds and inconsistent Internet service provider quality, a potentially vast market demand exists for wireless computer network dial-up. Cheap, ubiquitous wireless access would break the bottleneck that the local exchange telephone company—and its narrowband copper wires—currently forms, and would obviate the very expensive, politically complicated, and economically counterproductive mechanisms now in place to enforce "universal service" and "wiring the schools" mandates.

Note that such a momentous social advance—wiring America via wireless—might inspire little in the way of auction bids for licenses. Suppose, for instance, that five national "overlay rights" (each exclusively assigned 80.4 MHz in bandwidth, for a total of 402 MHz) are to be sold via competitive bidding. Suppose further that 30 firms register to bid and that they do not collude (thus eliminating any strategic behavior and ensuring that the government receives the expected value of the licenses). Still further, assume that each of the firms projects that wireless Internet access service is the highest valued use for the frequencies and that each values the chance to compete (i.e., the value of a license) within a narrow range wherein revenues will just exceed costs. The top five (winning) bids, then, would presumably produce only modest receipts for the Treasury. Would this "disappointing" FCC auction indicate a policy failure?

Quite the reverse. The auction would, in fact, constitute a grand success. Regulators would have produced so competitive an equilibrium that license rents would be trivial. Consumers, on the other hand, would gain the enormous benefit of high-speed wireless service offered to millions of U.S. customers at prices *so low* that the operating licenses are worth very little. At bottom, lowering license values through pro-competitive reforms should be the goal of public policy. There is no doubt that there is much pent-up demand for bandwidth, and there is simply no social opportunity cost in unleashing wireless suppliers to provide it.

Political Obstacles to Reform

One of the most important recent developments in telecommunications regulation is the eclipsing of broadcast interests at the FCC. Until the emergence of cellular telephone service, broadcasting was virtually the sole concern at the FCC on the wireless side. Only with the advent of high-ticket licensing issues in nonbroadcast markets (specifically, cellular and PCS) did Congress relax its insistence on political assignment of licenses (via comparative hearings) and allow more neutral mechanisms (lotteries in 1981, just before cellular; auctions in 1993, just before PCS).

Broadcasting evokes a heightened level of interest among policymakers precisely because the industry's output—publicity—is simultaneously an input into the production function of politicians. This situation creates the opportunity for in-kind trades, provided the officeholder is in a position to bargain. The desire to control radio program content is what steered lawmakers to enact the vague "public interest" spectrum policy suggestion of broadcasters. Broadcasters happily traded away a bit of editorial control in exchange for favored positions in the marketplace. In return, they received free licenses and protection from new competitive entry—a combination package that the "public interest" system delivered and its predecessor (the common law rules of "priority in use") could not.

The regulatory scheme concocted in 1927 brought radio regulators and radio broadcasters into a quid pro quo arrangement beneficial to both sides. Regulators intensely desired to shape content decisions, as the growing influence of the new medium of communication was clearly felt. Major commercial broadcasters wanted both protection of their existing interests in radio broadcast frequencies and, if they could get it, a denial of the claims of newcomers. The "public trusteeship" concept enacted effectively granted broadcasters a super-property right in broadcasting licenses: not only did incumbents have the right to access radio waves without fee, but regulators would force potential entrants to prove that competition was in the "public interest" before allowing rivals access to the broadcasting market.

In the seven short years in which the Federal Radio Commission operated (1927–34), the major commercial radio interests enjoyed substantial windfalls. Rules crafted by the commission decimated the ranks of both nonprofit stations (owned by churches, schools, universities, local governments, labor unions, social organizations, etc.) and smaller, less profitable commercial outlets. The reduction amounted to about 200 of the approximately 750 radio stations originally operating. Curiously, this "cleansing" of the radio market followed the commission's 1927 decision *not* to expand the AM dial to accommodate the "surplus" stations, at a time when large, network-affiliated concerns were receiving enhanced power assignments. The financially important radio incumbents could not have devised a more profitable outcome.

Only by arguing that the "public owns the airwaves" could lawmakers create this one-way property right. Hence, under U.S. law, neither private parties nor the government owns the airwaves; rather, the public does. The government, pursuant to the 1927 Radio Act and the 1934 Communications Act, assumes responsibility for policing access to the public's property so that it is not rendered worthless via conflicting transmissions.

Similarly, the recent debate over high-definition television included a barrage of broadcaster lobbying about the rights of every American to "free, over-the-air TV." The asserted public right is the broadcaster's premise for the conclusion that the government should preempt any market test of the public's allegiance to "free, over-the-air TV." In reality, of course, the bands walled off for "free TV" cost the American public dearly.

The alternative uses of the TV band are self-evidently more compelling in their public good components. Consider the choices: Protect "free" TV for the 30 percent of households without cable or satellite or build a ubiquitous wireless telephone network so cheap that low-income people could afford it? Sharper picture resolution for "Spin City" and "ER" or the crime deterrence effects of a 90 percent cellular telephone penetration rate? Safeguard the educational uses of broadcast television or rescue would-be providers of high-speed Internet access from spectrum starvation?

The arguments defending the extant regime constitute no argument at all; the words merely spin cover stories, unsubstantiated tales—but that is sufficient in the zero-scrutiny world in which sound-bite justifications entitle members of Congress, commissioners, and presidents to regulate according to their political self-interest. There is no discernible substance to the arguments advanced in the spectrum skirmishes; what counts is the process.

Even when market incumbents are shooting "public interest" blanks, their obfuscations and scare tactics can strategically manipulate the political and administrative channels by which spectrum allocation issues are refereed in the "public interest." That underscores the importance of systemic reform: better arguments and keener analyses within the existing framework will not likely bring cleaner results or rulings more favorable to the interests of consumers, voters, or taxpayers.

9. The Effect of Privatization on Free Television

Stanley S. Hubbard

I would like to give you my opinion about ownership rights in spectrum insofar as radio and television are concerned. For purposes of this paper, I will stick to television.

But first, please understand that I, my family, and the company my associates and family operate always act in accordance with the letter and the spirit of the law. Although we may not agree with some particular laws or rules that govern our country, we nevertheless always work within the framework of our legal and moral system. We operate our radio and television stations in accordance with the spirit of the law and the law itself, insofar as the Federal Communications Commission, other administrative agencies, and the Congress have established the rules and policies governing the operation of broadcast stations in the public interest, convenience, and necessity.

When the FCC first made television construction permits available, it was extremely difficult to find individuals or companies willing to make the investment required to build a television station. U.S. television first began experimentally in 1939, and full-service commercial stations were licensed in 1947. At that time, only a handful of television stations were on the air. Those television stations were, for the most part, owned by large corporations, such as RCA, Westinghouse, or DuPont.

A good example of the difficulties encountered in the early development of free, over-the-air television is what occurred in my own family. My father, Stanley E. Hubbard, was a radio pioneer and, later, a television pioneer. In 1939 my dad purchased from RCA the first television camera that was sold in the United States. With that camera, in the summer of 1939, he telecast an American Legion

Stanley S. Hubbard is president and chief executive officer, Hubbard Broadcasting, Inc.

parade in downtown Minneapolis, Minnesota. The single television camera pickup, with narration by radio personality Rock Ulmer, was fed via closed circuit to six television sets in the old downtown Radisson Hotel. People could go into the hotel to view the live parade on six RCA reflective mirror television sets.

At that time, my dad was one of the few pioneers who truly believed in television's potential to benefit the American people. He wanted to start his television station, KSTP-TV, in St. Paul-Minneapolis. The problem was that he lacked the capital; RCA was not yet ready to begin the commercial production of television transmitting equipment and receivers, and the FCC was not accepting applications for television construction permits. World War II started in 1941 and all plans for the development of television were put on hold.

In 1947 the green light was given to proceed with the launch of commercial television service in the United States. Once again, my dad faced the money problem. How could he get the capital to build his dream, KSTP-TV? The government certainly was not going to be of assistance, and it was not the place of government to help. Private investors and bankers were not helpful because they could not visualize the potential of the television medium. My dad went everywhere he could think of to try to raise money to build the television station and, at every turn, he met with rejection. It was difficult to get people to invest in television even though the FCC did what it could to promote the development of the medium.

I remember attending a meeting that occurred as a result of the then-chairman of the FCC, Lawrence Fly, having asked my dad to convene a group of broadcasters from North and South Dakota, western Wisconsin, Iowa, and Minnesota. The purpose of that meeting was to provide an opportunity for the chairman and my dad to try to convince the radio broadcasters to invest in building television stations. I was at that meeting and it was a hard sell.

In the early days you could almost have received a television license for the price of a postage stamp. Large and successful companies frowned on television. Commander Eugene McDonald, who was the chief executive officer of the Zenith Corporation, returned his company's license for cancellation. That authorization later became Channel Two in Chicago. CBS held a meeting with its radio affiliates in 1947 at which CBS chairman Bill Paley and CBS president Frank Stanton advised their radio affiliates not to invest in television because FM was going to precede television and, they claimed, television would not take off for quite some time thereafter. However, people like RCA-NBC chairman General David Sarnoff and my dad persisted. Eventually my dad was able to arrange financing with the Mellon Bank in Pittsburgh to build his television station. To do so, however, he had to mortgage his radio station, KSTP-AM.

The point of this story is that no one helped my dad. He had to do it himself. He put everything he had at risk. The government did not grant a license with a guarantee of success. As a matter of fact, any student of the history of television will find that at that time most of the so-called experts in media were predicting that television would be no more than a passing fad and those who invested in it would fail.

The Broadcasting Spectrum Has No Intrinsic Value

What was the value of the spectrum in 1948? Nothing! The government had trouble giving it away. I remember attending a radio convention in Chicago where my dad was trying to convince radio operators to enter into television. Some of the radio operators—a few of whom later became bitter old men—said that they considered my dad a traitor to radio and that he would "lose his shirt" if he invested in television.

The point is this: The television spectrum initially had no value. None! No one wanted it. The government sought to talk people into building television stations. Those stations, of course, would use the spectrum. The television spectrum did not have value in and of itself, although it was available. The value of the spectrum resulted from risky investments, hard work, and creative entrepreneurship. We, the television broadcasters, made the spectrum for television broadcasting valuable. Like those who pioneered the early West and were given homestead rights, the television pioneers who created the value in the spectrum should have free and clear title to that spectrum.

The same reasoning is applicable to the satellite broadcasting spectrum. From 1981 until the 1990s, my family and our associates tried to interest people in investing in the development of a Direct Broadcast Satellite (DBS) system. It was not until we were able to enter into ownership of a high-powered satellite with Hughes Communications Galaxy, Inc., that we were able to interest a few farsighted investors in joining us in starting a high-power DBS service. What was the DBS spectrum worth before we demonstrated a use for it? Nothing! Once again, it had no value. Through our risk investment, we created value in the DBS spectrum where there had been none. It is interesting to note that in the mid-1980s, the FCC even considered deleting the DBS spectrum from the allocation table.

I, for one, have no problem with spectrum auctions for those who follow the pioneers who have established a real spectrum value as a result of the service they have provided. For example, the FCC held an auction and took bids for the DBS spectrum at the 110° west longitude orbital position. That may have been appropriate because the value of the use of that spectrum for DBS had already been proven and established by those of us who took all the risks necessary to develop DBS in this country. Just as with terrestrial television, I believe that those who took all the risks in pioneering the development of the spectrum—like the early pioneers who opened the West—should be entitled to keep it. The spectrum in and of itself is nothing but ether. It is because of risk investment that the spectrum gains value. Those who put the capital in and took all the risks should be entitled to lay claim to the spectrum, the use for which they have developed, whether it is local, over-the-air broadcasting, or DBS.

The Airwaves Differ from Other Natural Resources

If the airwaves are to be considered a natural resource, then it should be realized that they differ from any other natural resources, such as minerals, oil, or gas. The airways differ in two principal ways: First, the airwaves are just that—the air. There is nothing to deplete; nothing is injured. People create an electronic signal that is sent through the ether, and the ether is merely the ocean upon which the signal travels. The signals, which are created by humans, are, when in use, called the airwaves. In the case of oil or minerals, there is immediately something of intrinsic value: oil, gold, natural gas, silver, copper, diamonds, or whatever. People do not create mineral or oil resources, but people do create the signal in broadcasting.

Second, and very important, mineral and oil resources differ from the airwaves in that by using the airwaves, people do not deplete the supply. Every hour broadcast does not represent a depletion of the remaining spectrum; whereas every ton of iron ore mined or every barrel of oil pumped from the earth depletes the amount remaining for future use. The airwaves are an intangible resource, and the signals transmitted on the airwaves are not dependent on some depleting intrinsic characteristic of the airwaves themselves.

When I was a boy, no one dreamed that we would be broadcasting today on the 12-gigahertz band. In fact, at that time, people were just beginning to experiment with the VHF (very high frequency) band. Who knows what frequency band may be used in the future? The airwaves are infinite. The airwaves can be expanded and the signals transmitted can be squeezed.

Broadcasters Should Have Clear Title to Their Licenses

If you look at the value of anything—for example, land—at what point did that land belong to the public? If you trace the origin of the land upon which thousands of frontier ranches were formed, you must ask yourself this question. Did all the developers of all those properties buy that land from the government, or were some of them given title to it in exchange for agreeing to develop it? When farmers went to Minnesota or to Montana and were given land by the government, did those farmers then not retain clear title to that land—clear title in return for developing that land and thereby expanding the vitality and the economic base of our nation?

The transcontinental railroads were built because America needed a transportation system. The government therefore agreed to give land to companies that would build the railroads. The Great Northern Railroad, for example, was given every other square mile of land between St. Paul and Seattle. That is an example of the kind of property right that should have been awarded in the early development of radio and television. Broadcasters should be treated the same as railroad builders or oil prospectors who laid claim to land in Oklahoma and Texas and were given the land as an inducement to prospect for oil. Broadcasters "staked a claim" in the early days of radio, for example, by building transmitters and putting stations on the air. Later came television. The only legitimate function of government in that regard should have been the allocation and protection of those frequency rights—protection for pioneer broadcasters from interference—just as the government guaranteed protection against trespass or land infringement to those who homesteaded in the early days of our country's development.

The adoption of the public interest, convenience, and necessity as the basis for awarding of radio and television station licenses was largely political! The basis upon which licenses were granted did not increase the efficiency of the radio system. It probably hurt efficiency, but it did provide politicians, for the first time, the opportunity to exercise control over a segment of the communications industry.

In my opinion, a free and open competitive market best benefits the public interest. Those individuals and companies that pioneered and own radio and television stations should have clear title to their licenses just as if the licenses were land grants. The American public is intelligent and knows what it expects from radio and television. Radio and television service, like any other business, will succeed only to the extent that it serves the public interest as public listenership and viewership define it. However, I do see a legitimate need to have government rules that place limitations on the amount of spectrum any one company may have.

The government should avoid micromanaging radio or television and limit itself strictly to protecting property and otherwise enforcing the laws of the land as they apply to newspapers, magazines, and other media. Following such a policy would greatly enhance the value of the use of the spectrum and work to the public benefit. Broadcasters and other developers of the use of the spectrum should be free to do whatever they wish within the confines of the general laws of the land. Such a policy, in my opinion, would maximize further development of productive use of the spectrum.

10. Free the Spectrum: Market-Based Spectrum Management

Evan R. Kwerel and John R. Williams

Introduction

Policies that lead to efficient spectrum use will help to promote competition, stimulate investment in infrastructure, and create jobs. Much of the attention generated by the 1996 broadband Personal Communication Services (PCS) auction focused on the revenue raised. Yet the primary benefit of the allocation of spectrum for PCS and subsequent auctions was to make more spectrum quickly available to the private sector.

The private sector will generally maximize social welfare by putting spectrum to its highest value uses. Spectrum rights can be allocated to their highest value use by a market only if they are available to the market, that is, exclusively and exhaustively assigned and freely transferable. The traditional system in which the government allocates bands of spectrum to narrowly defined uses and retains large amounts of unassigned spectrum largely prevents the operation of an efficient spectrum market. The Federal Communications Commission has made much progress in incrementally implementing a market-based spectrum policy, although much remains to be done. In the last five years it has made large blocks of spectrum available for new uses, defined spectrum usage rights much more flexibly than in the past, and generally allowed free transferability of licenses. Further rapid progress toward the market allocation of spectrum depends on the commission's success in addressing difficult policy and political issues, including finding efficient solutions to potential market failures, providing fair and efficient treatment

The authors work at the Office of Plans and Policy, Federal Communications Commission. The opinions and conclusions expressed in this paper are those of the authors and do not necessarily reflect the views of the FCC, any of the commissioners, or other staff.

of spectrum incumbents, and maintaining support for rapidly auctioning additional spectrum as prices fall.

The Case for Market Allocation of Spectrum

Many economists have argued that the FCC should adopt a market-oriented approach to the allocation of spectrum.[1] Economic theory and historical evidence suggest that allocating scarce resources by markets will increase consumer welfare more than doing so by government planning. If the FCC's principal objective is to allocate spectrum efficiently, and if spectrum markets can be structured to avoid significant market failures, then markets should be the primary method for allocating spectrum. Assigning most spectrum usage rights, making those rights exclusive and freely transferable, and auctioning licenses will minimize the possibility of market failure.

Even if specific market failures or conflicting social objectives can be identified with regard to certain spectrum uses, one cannot automatically conclude that government allocation for such uses will be superior to market allocation. Government allocation is subject to its own distortions, and "government failure" may cause a nonmarket allocation to be inferior even to a less-than-optimal market allocation.

Nevertheless, if specific market failures or conflicting social objectives can be identified, limited and carefully targeted regulatory intervention may be appropriate. One advantage of a policy that relies on markets as the primary method for allocating spectrum is that it generally makes explicit the opportunity cost of pursuing social or political objectives that conflict with the goal of allocating spectrum efficiently.

Framework for an Efficient Spectrum Market

Spectrum Rights Should Be Exhaustively Assigned

If markets are to award spectrum to the parties that are willing to pay the most (and thus presumably will put it to its highest valued use), the rights to that spectrum must be assigned to an identifiable party that is free to use or trade them. Those rights include all uses at all points in time, frequency, and geographic location.

Spectrum licenses should be defined by geographic areas and spectrum blocks that exhaustively cover the United States and the available spectrum. That is in contrast to the traditional practice of licensing specific frequencies at specific sites or groups of sites as is done, for example, for broadcasting and point-to-point microwave services. The FCC or National Telecommunications and Information Administration (NTIA) should not generally hold spectrum in reserve for later assignment. The government should rely on the market to provide appropriate spectrum "reserves," just as private developers of land and other natural resources hold reserves in anticipation of future demand.

Licensees should be given full flexibility in the use of their spectrum. Permitting licensees to determine spectrum use would provide large benefits by allowing the market to shift spectrum to higher value uses in response to changing technologies and consumer demands. There should be no set-aside for any category of use except to correct for specific market failures, such as the possible underprovision of low-power devices. To allow for market determination of spectrum uses, it would be necessary to permit broad flexibility of use for both new and existing licenses and to license blocks of spectrum of sufficient size and geographic scope to permit multiple uses. It would no longer be necessary or appropriate for the government to evaluate competing claims of would-be service providers submitted in a rulemaking proceeding.

Licensees should be allowed to employ any technology subject only to interference constraints at the geographic and spectrum boundaries defined by their licenses. In a well-functioning spectrum market, licensees face the full opportunity cost of spectrum use and make the efficient tradeoffs between spectrum-conserving technology and additional spectrum use. Fortunately, the commission has been moving toward allowing technological flexibility in its recent allocations.

Buildout schedules should be determined by markets. Having the opportunity to sell their licenses or change the use of their spectrum at any time means that, in general, licensees would face the full social cost of holding spectrum for future use. Performance requirements thus would not increase efficiency and could cause licensees to build uneconomic facilities merely to satisfy the terms of the license. Performance requirements have sometimes been justified as a means of preventing warehousing by firms with market power.

However, if spectrum use were broadly deregulated, firms would be unlikely to be able to warehouse sufficient spectrum to prevent the entry of competitors. Rivals would have strong profit incentives to provide services that earned above-market rates of return and, without regulatory barriers, would be able to do so by shifting spectrum from other uses. In the short run, however, there may be transaction costs that prevent spectrum from being rapidly deployed in a particular service. Therefore, it may be desirable to place a temporary cap on the amount of spectrum suitable for specific services that any one firm can control in an area.

Licenses should be awarded in perpetuity. Such a policy encourages efficient investment in equipment and other factors such as marketing that may be tied to holding a specific license because it allows license holders to retain the benefits of investments. Without the ability to acquire rights in perpetuity (or negotiate effectively about investments tied to a specific license), the current licensee would tend to underinvest in license-specific factors, especially as the end of the license period approached. Although licenses currently are awarded for fixed terms of up to 10 years, the high renewal rate creates de facto licenses in perpetuity.

Initial licenses should be awarded by auction. Among the advantages of auctions over awarding licenses by comparative hearings or lotteries are (1) promoting economic efficiency by awarding licenses to those who value them the most, (2) reducing wasteful private expenditures (rent seeking) on obtaining licenses, (3) minimizing delays in licensing, and (4) raising revenue. The benefits would exist for any service for which mutually exclusive applications are filed.

Spectrum Usage Rights Generally Should Be Licensed Exclusively

The basic problem with assigning shared spectrum is that spectrum use by an individual licensee creates an externality for the other licensees that share the same spectrum. Individual users of shared spectrum have little incentive to conserve on the use of the spectrum because they cannot capture for themselves the full benefits of their conservation efforts, creating a "tragedy of the commons" problem. That has been the FCC's experience with traditional shared land mobile services, which are characterized by an excessive number of mobile units, each providing low-quality service. In contrast, the FCC has found that specialized mobile and cellular radio operators, who are awarded spectrum on an exclusive basis, provide a range of service quality and voluntarily use spectrum-conserving equipment.

Under a market-based system, the FCC would define and enforce exclusive spectrum licenses. The only significant exception would be low-power devices that would be permitted to operate unlicensed because of their low potential for interference. Implementing market-based allocation in shared spectrum will require phasing out the shared-use licenses and replacing them with exclusive licenses. Of course, the exclusive licensees of such spectrum could still use their spectrum to create systems that are technically the same as shared use systems, but they would also have the right to limit entry and otherwise manage spectrum use and would realize the opportunity cost of the spectrum use.

Spectrum Usage Rights Should Be Freely Transferable

If users cannot freely transfer among themselves all categories of spectrum usage rights, markets will not be able to put such rights to their highest value uses. Licenses are now generally transferable within service categories, but not all parties are eligible to hold licenses. Under the market-based approach, parties would be free to transfer, subdivide, and aggregate flexible-use licenses in all dimensions (time, frequency, and geography), subject only to temporary spectrum caps to prevent excessive concentration in geographic markets. That would include allowing transfer of spectrum between government and nongovernment users. If subdividing licenses imposes significant additional costs on the FCC for enforcing interference rights at the newly created license boundaries, it may be necessary to establish a subdivision fee schedule to provide private parties with the incentive to subdivide efficiently.

Progress toward a Market-Based Spectrum Policy

The FCC has made much progress toward implementing a market-based spectrum policy. Most important, it has put large blocks of new spectrum in private hands. Since July 1994 the FCC has conducted 14 auctions of more than 4,000 licenses for exclusive usage rights. Much of that spectrum, including the 120 MHz allocated for PCS, was made available by exhaustively licensing partially occupied bands. New licensees were allowed to use any available spectrum within a specified spectrum block and geographic area consistent with protecting incumbent licensees from interference. More spectrum is to be auctioned soon, for example, Local Multipoint Distribution Services (LMDS) with over 1,000 MHz (in 28-GHz range). Additional blocks have been proposed in the so-called millimeter wave bands above 30 GHz. The FCC also made large amounts of spectrum available on an unlicensed basis (nonexclusive), including 300 MHz of bandwidth in the 5-GHz range (the U-NII band) and a large block in the 60-GHz range.

Second, broad service and technical flexibility were permitted for most of the newly available spectrum. For example, the definition of PCS in the 2-GHz range allows virtually any mobile or fixed point-to-point uses, although broadcasting and satellite services are not permitted. Most of the new or proposed bands to be auctioned in the future also will provide wide flexibility to licensees to choose the services they provide. Still, licensee control over service selection is the exception rather than the rule in most of the spectrum. In the UHF (ultra high frequency) range from 300 MHz to 3,000 MHz, for example, the PCS and cellular bands account for only about 6 percent of total bandwidth. In the rest, licensees are limited to single, prescribed services such as television broadcasting, private land mobile, or fixed point-to-point. Also, in many of the bands, special provisions have been made for federal government use that either preclude or limit spectrum availability for the private sector.

Third, while flexibility of use is limited in most bands, the licenses themselves are generally transferable, which allows market allocation of spectrum among users within the defined service category. The main exceptions are licenses auctioned with preferences such as bidding credits or set-asides. Such licenses cannot be transferred to users outside the eligible class for a fixed number of years.

Progress toward a market-oriented spectrum policy has not been universal, however. The DTV (digital television) reallocation is probably the most economically significant recent exception. Congress explicitly rejected the use of auctions to assign DTV licenses. The FCC decided to manage the transition to DTV administratively, instead of relying on a market approach analogous to that employed for PCS, which would have called for exhaustively licensing all the spectrum in the television bands and letting the private sector determine the timing and nature of the transition to higher value uses of the spectrum.

Policy Transition Issues in Implementing Market Spectrum Management

Creating new property rights and changing old ones to permit spectrum markets to operate efficiently is not a costless process. Even if we were starting with virgin spectrum, it would be necessary to define initial license boundaries—frequency block size, geographic scope, interference—and to determine an auction schedule. The initial auctions and the aftermarket can adjust license boundaries, but not without cost. The FCC may be able to reduce such transaction costs by setting the initial boundaries according to expected uses. The auction schedule is a key policy decision as well. Resource constraints and legal requirements for notice and comment prevent the FCC from auctioning all available spectrum immediately. But even if it were possible, it might not be optimal. The private sector might need time to form a consensus on likely uses, coordinate among manufacturers and service providers, and raise capital. Failure to provide sufficient time could lead to an inefficient initial distribution of spectrum and subsequent high transaction costs. That may partially explain the low revenues from auctioning WCS (Wireless Communications Service) spectrum under the compressed schedule required to meet congressional deadlines.

Additional policy issues arise when, as is generally the case, the FCC is licensing spectrum that is partially occupied. The distribution of rights to remain or move, with or without compensation, matters both for creating political support for reallocating the spectrum to new uses and, once a reallocation is made, for minimizing bargaining costs of achieving an efficient reallocation.[2]

Possible Market Failures in Implementing Market Spectrum Management

Broadcasting

Broadcasting services have a key characteristic of a pure public good—nonrivalry in consumption—because adding users to a broadcasting service does not detract from consumption by others. The social value of advertiser-supported ("free") broadcasting includes both the value to advertisers and the value to viewers. However, broadcasters' demand for spectrum accounts for only the value to advertisers. Thus, in the absence of government intervention, a spectrum market may allocate less than the socially optimal amount of spectrum (as well as other inputs) to broadcasting. That may justify a direct subsidy, as is provided to public broadcasting, or setting aside certain spectrum for broadcasting only. In practice, however, determining the appropriate subsidy or set-aside for broadcasting proves difficult, and there is no reason to believe that the amount of spectrum currently set aside for broadcasting is efficient. Further, consumers almost certainly enjoy a significant surplus (benefits in excess of what they pay) from nonbroadcasting uses of the spectrum such as cellular service. An unregulated market thus may provide either too little or too much broadcasting relative to other services.

Public Safety and National Defense

Public safety and national defense users have traditionally argued for special treatment because they protect human life and property. Once most spectrum is available for market allocation, there is no reason that spectrum inputs should be treated differently than are any other inputs (e.g., vehicles, ammunition, and radio hardware) that also are used in the preservation of life and property. Indeed, requiring public safety and national defense users to pay for their spectrum (and allowing them to keep the funds from selling some of their current spectrum) would result in more efficient use of the spectrum because such users would be forced to consider the social cost of using spectrum in place of other inputs.

Unlicensed Use

Unlicensed use may be justified for a limited amount of spectrum because such use has the characteristics of a public good. It may not be efficient to license or charge for entry if one person's use does not significantly contend with another's. That arrangement normally requires sharing protocols and accepting severe limitations on transmitter power and range. Because such restrictions preclude valuable licensed uses, determining the appropriate amount of spectrum to allocate for unlicensed use would require a cost-benefit analysis. In practice, administrative agencies find it difficult to conduct such analyses and will end up assigning an arbitrary amount of spectrum to such uses. Further, holders of exclusive spectrum licenses may be able to provide similar services by charging a license fee to manufacturers of low-power devices that operate on their spectrum or by limiting use of the spectrum to low-power devices that they manufacture. Although the process may result in a suboptimal allocation of spectrum, that allocation will not necessarily be worse than an administratively determined one.

Standards

A pure market may fail to set standards efficiently because of the high transactions cost of negotiating standards and the public good aspects of standards (i.e., once some parties have borne the cost of setting a standard, other parties can benefit from adopting it at no additional cost). Socially optimal standards also may not develop in a pure market because dominant providers may strategically refuse to establish open standards in order to disadvantage rivals. That fact does not necessarily imply that government involvement in standard setting is appropriate, however, as standard setting by administrative process tends to create rigid, and often wrong, standards after costly delays.

International Services

International services, almost by nature, require use of spectrum in more than one country. There are political and institutional obstacles to worldwide spectrum auctions, including disagreement on who would run the international auction and on how to divide the proceeds. Those difficulties do not, however, prevent the United States from using auctions domestically and relying on both government and private negotiations to acquire licenses abroad.

Social or Distributional Issues

The 1993 legislation authorizing auctions contains explicit provisions requiring the FCC to ensure that small businesses and businesses owned by women and minorities "are given the opportunity to participate in the provision of spectrum-based services."[3] The commission used bidding credits, set-asides, and installment payments to implement that legislative mandate. One possible market failure rationale for the use of installment payments is capital market imperfections. Capital market imperfections may cause some firms to have difficulty attracting capital because of discrimination or information asymmetries.[4] Information asymmetries could be a particular problem when lending to new entrants with a limited credit history. We have no empirical evidence that this has resulted in market failure, however; and the FCC's program to help small businesses by selling licenses on an installment basis did nothing to directly alleviate any information asymmetries. Moreover, the program may have helped cause the high default rate in the PCS C block auction. Because of limited liability, firms with the riskiest business plans may have been willing to pay the most for loans that the FCC offered in conjunction with each license, thus biasing license assignments toward firms with the highest risk of default.

The 1993 auction legislation also requires the commission to provide safeguards to facilitate the delivery of new services to rural areas. Such universal service objectives do not necessarily conflict with using markets to allocate spectrum. Indeed, greater reliance on market forces will promote universal service by lowering costs and prices to consumers. To the extent that some consumers are still unable to purchase some services that are considered socially essential, direct subsidies to low-income consumers are likely to be the least costly way to expand consumption.

Conclusion

A carefully designed system of market-based spectrum management would permit much more efficient spectrum allocation and significantly increase consumer welfare. A critical element of that approach is making more spectrum available to the market with minimal regulatory constraints on use, as the market cannot efficiently allocate what it does not have. Market failure issues that cannot be addressed through system design, and goals other than efficient resource allocation, can be addressed by relatively small and targeted limitations on the scope of market-based spectrum management. The broad and flexible service rules, the innovative band-clearing plan, and the successful use of auctions to assign licenses in PCS provide the necessary elements. It is time to build on PCS experience and apply the techniques introduced there to the rest of the frequency spectrum.

Notes

The authors would like to thank Greg Rosston and Michael Katz for their many helpful comments.

1. See Ronald H. Coase, "The Federal Communications Commission," *Journal of Law & Economics* 2 (1959): 1–40; Arthur Devany et al., "A Property System for Market Allocation of the Electromagnetic Spectrum: A Legal-Economic-Engineering Study,"

Stanford Law Review 21 (1969): 1499–1561; Jora R. Minasian, "Property Rights in Radiation: An Alternative Approach to Radio Frequency Allocation," *Journal of Law & Economics* 18 (1975): 221–72; Douglas W. Webbink, "Radio Licenses and Frequency Spectrum Use Property Rights," *Communications & the Law* 9 (1987): 3.

2. See Peter Cramton, Evan Kwerel, and John Williams, "Efficient Relocation of Spectrum Incumbents," Paper presented at Telecommunications Policy Conference, Solomons Island, Md., October 1996.

3. 47 U.S.C. § 309(j)(4)(D)(1997).

4. Joseph E. Stiglitz and Andrew Weiss, "Credit Rationing and Markets with Imperfect Information," *American Economic Review* 71 (1981): 393–410.

11. Beyond Auctions: Open Spectrum Access

Eli M. Noam

Three Old Paradigms and a New One

It will not be long, historically speaking, before spectrum auctions may become technologically obsolete, economically inefficient, and legally unconstitutional.

And it may not be long before a new form of frequency allocation emerges wherein spectrum use does not require a license; information traverses the ether as flexibly as an airplane in the sky instead of being straitjacketed into a single frequency and routed like a train on a track; and congestion is avoided not by the exclusivity of ownership but by access charges that vary with congestion, with the information itself often paying for access with tokens it carries along.

For today, auctions and usage flexibility still offer the best way to allocate new frequencies. Yet it is one thing to support them pragmatically, as I do, because they present a better approach than the existing alternatives, and quite another thing to regard auctions with dogmatic awe, blind to their technological relativism. Change the technology and the economics and the law of spectrum use must change, too.

The *auction* paradigm has more in common with the *regulatory assignment* paradigm it threatens to supplant than the proponents of either care to admit. Both basically allocate exclusive slices of the spectrum and differ only in the initial mechanics of that allocation. Seen thus, the two paradigms really collapse into a single one, that of *licensed exclusivity*.

But now new technologies, available or emerging, make possible new ways of thinking about spectrum use. The new paradigm is that of *open access*, under which many users of various radio-based

Eli M. Noam is a professor of finance and economics, Columbia University Graduate School of Business, and director of the Columbia Institute for Tele-Information.

applications can enter spectrum bands without an exclusive license to any slice of spectrum, by buying access tickets whose price varies with congestion. The tickets could be carried by the information itself. That would take us back, in several ways, to the earliest stage of frequency use, when there were no licenses. That revolution will become an option soon because we will be able to solve the problem of interference that doomed the occupancy model and led to the licensing system in the first place.

The rumblings against the auction paradigm emerged in the mid-1990s from Paul Baran and George Gilder.[1] Underlying their arguments is the hope that technology will solve the scarcity problem and spare much of the need to deal with allocation questions.[2] In contrast, I argue that in an open access system scarcity emerges, the resource needs to be allocated, and a price mechanism can do it.[3] But that does not require exclusive control over a slice of the spectrum.

Spectrum policy is traffic control, not real estate development. The traditional spectrum approach can be compared with the granting of a limited set of taxi medallions for free, renewing them nearly automatically, permitting an aftermarket, and restricting the issuance of new licenses to protect their value. The analogy fairly exactly describes the notorious taxi-licensing system of New York City, where rates keep going up in order to assure buyers of medallions a proper return, thereby further raising the price of the medallions, and so forth. An auction system would merely change the initial mechanism of distributing medallions. Instituting flexibility of usage would allow a medallion holder also to provide, say, ambulance service. But open access would permit any firm to enter taxi services, without license or usage limitation. To remedy congestion of bridges or major roads, drivers would pay at tollgates with tokens the price of which was set automatically to clear congestion.

Whose Spectrum Is It Anyway?

The open access system offers more than just increased efficiency in technology, economics, and policy. More important, it offers the prospect of a more *constitutionally sound* system for allocating rights to the spectrum. Because the First Amendment protects electronic as well as print speech, the state may abridge speech-via-spectrum only in pursuance of a "compelling state interest" and through the "least restrictive means" that "must be carefully tailored to achieve such interest."[4]

Spectrum licensing schemes look constitutionally suspect because they inevitably foreclose the electronic speech of those without licenses and limit the speech of those who have them. Until now, the government justified licensing on grounds that spectrum was a scarce resource that required careful protection against interference. But open access offers the possibility of abundant spectrum. The realization of that possibility would make it no more constitutional for the state to auction off exclusive rights to access spectrum than it would be for New York City to auction off exclusive rights to speak in the street.

Instead of loosening the barriers to free access, however, the U.S. government is going in the opposite direction, by selling off the spectrum. But is the spectrum the government's to sell in the first place? It is one thing to serve as a traffic cop, keeping the different users from colliding. It is quite another matter to assert ownership rights (in effect, to nationalize the spectrum retroactively) and to sell them off. If electronic communications are an aspect of our fundamental free speech rights, on what ground can those rights be sold to the highest bidder? Advocates of the free market do not usually grant governments the authority to appropriate the economic value of attractive commercial opportunities.

The Future Problems with Auctions

Today, almost everyone loves auctions: most leftist liberals, because auctions make business pay its way and generate government revenues; and most conservatives, because auctions substitute market mechanisms for government controls. Auctions have been used in New Zealand, the United States, the United Kingdom, Australia, Hungary, and India. Others countries will follow, no doubt. On the whole, the arguments in favor of auctions are stronger than the arguments against, partly because most legitimate problems raised by the critics can be dealt with in other and often more efficient ways. But auctions do have problems.

Auctions Inevitably Deteriorate into Revenue Tools

Auctions were finally approved, after years of opposition to them by powerful congressional barons, only as a measure to reduce the budget deficit while avoiding spending cuts and tax increases.

Allocating spectrum resources efficiently was a secondary goal. Conceived in the original sin of budget politics rather than communications policy, spectrum auctions are doomed to serve as collection tools first and allocation mechanisms second.

Several problems are tied to the budget-driven auction system. One is a spend-as-you-go approach. It is one thing to sell assets (spectrum rights) and reinvest the proceeds. But ours is a situation of funding current consumption through the sale of long-term assets. That exposes spectrum auctions to political dynamics. The 1996 Telecommunications Act created a Development Fund, aimed at small minority businesses, to be funded from the interest on auction bids. Vice President Al Gore wants auction revenues to finance the wiring of schools. Rep. Jack Fields (R-Tex.) wanted to use them for public television. President Bill Clinton, for school rehabilitation. As this takes place, stakeholder groups emerge and seek ongoing funding and therefore ongoing auctions.

Around the world, countries aim to advance their national infrastructures. In the United States, there seems to be widespread agreement that it should be done without government money. But the spectrum sales end up serving the opposite end. Through auctions, we are taking money away from infrastructure-providing telecommunications firms and throwing it into the hole of the federal budget. For decades, the U.S. telecommunications system was superior to that of any other country, often because other countries used telecommunications as a cash cow for general government expenses. Now we have embarked on the same path, just when other countries have left it at our urging.

When all is said and done, an auction is a tax on the communications sector and its users.[5] Advocates of auctions deny that, arguing that consumer pricing depends on marginal rather than historic cost, and that the auction charge does not necessarily mean higher end-user prices if demand is highly elastic or if the rents have previously been squeezed by government in other ways. But how can anyone deny that the many billions of dollars raised by an auction are taken out of the private sector and end up with the government? That, after all, is the congressionally mandated point to the whole exercise.

Auctions Encourage Oligopoly

The highest bidders will be those who can organize an oligopoly. The arrangement is facilitated by bidding consortia of companies that otherwise would be each other's natural competitors and who collaborate under some rationale of synergy.

Oligopoly can be attacked in several ways: by adding spectrum allocations, encouraging spectrum flexibility, imposing structural rules of ownership limitation, and using antitrust law. That is indeed Federal Communications Commission policy. However, ownership limitations are regulatory in nature, may conflict with potential efficiencies of scale, and are in tension with the stated goal of moving spectrum to the highest valuing user.

Enough spectrum also must be auctioned off to attack oligopolistic tendencies and reduce opportunity cost. But here, government forces a conflict. Release more spectrum, and its price drops.

A Better Alternative: Open Spectrum Access

The alternative to the present auctions is not to return to the wasteful lotteries or comparative administrative hearings of the past but to take a further step forward, to full openness of entry, which becomes possible with fully digital communications. In those bands to which it would apply, no one would control any particular frequency. In that system no oligopoly can survive because anyone can enter at any time. There is no license and no up-front spectrum auction. Instead, all users of those spectrum bands pay an access fee that is continuously and automatically determined by demand and supply conditions at the time, that is, by the existing congestion in various frequency bands. The system is run by clearinghouses of users.

The underlying present auction system is premised on an analogy to land ownership (or long-term lease). It assumes a certain state of technology. In the past and present, the fixed nature of a frequency usage has had a stability that indeed resembled that of land. But that was based on the relatively simple state of technology; information was coded (modulated) onto a single carrier wave frequency or at most a narrow frequency range.

To forestall interference with other information encoded on the same carrier wave, the spectrum was sliced up, allocated to different types of uses, and assigned to different users. It is as if a highway were divided into wide lanes for each type of use—trucking, busing, touring, and so forth—and then further into narrow lanes, one for each transportation company. Once one accepts that model of spectrum, one can argue about how to distribute the lanes, whether by economics, politics, chance, priority, diversity, or other use. But it is important not to take the model as given and focus merely on optimizing it. To stay with the analogy, why not intermingle the traffic of multiple users? And if the highway begins to fill up, charge a toll to every user? And make the toll depend on the congestion, so that it is higher at rush hour than at midnight?

An Example of an Open-Access Model

The system I propose is not available using present technology, though its component parts exist or are within reach. I will not try to work out the details here, as they will evolve with time, discussion, and technology. Instead, I will offer a conceptual description of an open-access system.

To be transmitted, a packet of information would require an accompanying access token. Such a code would enable the packet to access a spectrum band, be retransmitted over physical network segments, and be received in equipment. Token prices would vary with congestion and be set according to demand and supply, the latter based on an initial endowment by an automated clearinghouse of spectrum users. A futures market could provide access at a price certain.

The buyer of tokens would own no particular slice of spectrum, but rather the right to send so many information packets at a particular time. At transmission time its equipment would scan for a free frequency. A receiver, similarly, would scan for information addressed to it. That is similar to the way local area computer networks work over wireline networks and now also over the air.

The clearinghouse could also auction off long-term access codes. In that case, it would approach the present auction and license system, except that no frequency exclusivity needs to exist, though that could also be instituted.

The access codes resemble tokens paid by drivers on toll roads or the tokens used in the "token ring" class of local area networks for computer data. With electronic cash emerging in the economy, the tokens could be general electronic currency. In effect, the information not only finds its own way (which packets already do) but also carries its own money for transit, picking among various over-the-air and wireline transmission options depending on price and performance.

Does the system require carriers? Wireline services obviously need their pathways to be maintained. But for over-the-air transmission, there is no roadway in the sky. Transmission firms more closely resemble airlines or shipping companies than railroad companies. They provide transmission and reception facilities[6] accessible by the information packets at a price. Those facilities need not permanently control any particular frequency any more than United Parcel Service and Federal Express control highway or air routes.

How to Implement an Open-Spectrum System

Who would administer such an open-access system? The options are (1) the government (but that would create powers of control and administrative inefficiencies that are undesirable); (2) the private owner of the spectrum (as discussed further hereafter); or (3) the users themselves, by way of a clearinghouse that functions like an exchange.

In practical terms, a clearinghouse would be a computer that sets access prices on the basis of demand. The resource it distributed would be the spectrum endowment that it controls. Access could be acquired in real time or in advance. Multiple clearinghouses for different bands are also possible and would provide competition.[7] The mechanism of a clearinghouse of providers has precedent in the way that the FCC has dealt with relocation issues in the PCS bands and in electricity distribution mechanisms.

Prices might be announced initially by a signal of spectrum price, based on supply and demand conditions, being sent out by the clearinghouse. There also could be different prices for different frequency bands, because their different propagation characteristics differentiate their attractiveness.

In special circumstances, a frequency could be dedicated entirely to a user or use, for example, to protect nonprofit, educational, or governmental use. Alternately, such users could receive a credit against which they could obtain access in the open-access system and which they could resell.

Who gets the proceeds? That is a political decision of allocation. The important point is that the revenue flow is smoothed, away from one-shot deals. Instead, the system converts fixed costs of entry into marginal costs of use. Transaction costs in an open-access system may be larger than in a traditional spectrum assignment system, but that is true for any open economic system. The offset is increased use and efficiency.

Early Examples of Open Spectrum Access

The FCC took early steps in the direction of open spectrum access in 1985 in its Part 15 rules, which increased power limits on the unlicensed use of spectrum bands used by industrial, scientific, and medical (ISM) low-power applications.

Progress toward open access continued in 1994 with unlicensed personal communications spectrum (U-PCS) opened to all users of asynchronous data and isochronous time-division duplex voice. Users coordinated spectrum access following a "spectrum etiquette" in real time, based on rules agreed upon by the industry. UTAM, Inc., a private nonprofit company owned by equipment manufacturers and supported by them in proportion to their U-PCS equipment sales, administers coordination, including the relocation of existing users and definition of channels and geographical regions.

The main weakness of the unlicensed access approach in its present stage is that it deals with scarcity and congestion through technological coordination. The best "etiquette" for the allocation of a scarce resource in our society is a market-clearing price. Without it one may reenact the rise and fall of citizens band radio, the poor person's open access, which suffers the absence of congestion prices and of commercial incentives for content provision.

Could an Auction Winner Administer an Open System Itself?

An auction winner administering its own open-access system would offer an alternative route to open access. If many wholesale spectrum band managers made resale transactions with many resale users, a substantial openness would indeed be achieved. But such a world seems unlikely. A monopoly or oligopoly is much more plausible. A wholesaler would need to control a significant band to provide meaningful access, which is likely to be unaffordable by any but the largest of telecommunications consortia. That is one lesson from the New Zealand experience. Alternatively, spectrum slices for wholesalers could be drawn narrowly, but then users would have less freedom to move around the spectrum and spectrum transactions among wholesalers would be controlled. Resale is clearly a step toward open access and should be encouraged. But although it is likely to exist in some limited fashion, it is not likely to generate a widespread openness of access.

Conclusion

The open-entry spectrum exchange will not solve every problem of today's auctions. New problems will emerge. Many of the problems may be resolvable once the technologists focus on them, but doing so requires first that we get out of the box of the exclusivity paradigm.

Even if the open-access system has flaws, the constitutional issue must still be answered. Efficiency of resource allocation and lower transaction costs do not overcome the protection of fundamental rights of which free (electronic) speech is one. If an open-access system is less restrictive than an auction-and-ownership model without causing spectrum chaos, the granting of exclusive speech rights may not pass the test of constitutionality. Even some inefficiencies and transaction costs cannot defeat constitutional rights.

What are some of the policy implications of open access? First, it should not stop auctions, since in the present state of technology they are still usually the better solution. But it should limit the duration of auctioned licenses in order to preserve future flexibility for other approaches.

Second, resale and spectrum use flexibility should be encouraged to facilitate resale markets. License holders should be able, in most cases, to slice up the spectrum and resell and sublet it to others for various applications.

Third, we should encourage experimentation and innovation. Why not, for example, expand the unlicensed spectrum concept and dedicate a few bands to the open-access, access-price model? Its practicality is a matter of technical evolution and market realities.

The proposed open-access paradigm is not likely to be accepted any time soon. But its time will surely come, and it will fully bring the invisible hand to the invisible resource.

Notes

The author has benefited from the research help of Po Yi, Nemo Semret, and Mobeen Khan, and from other assistance from Mae Flordeliza, Anne Behk, John E. Kollar, Carla Kraft, and Alex Wolfson. Bruce Egan, Robert Frieden, John Friedman, Thomas Hazlett, Wayne Jett, Evan Kwerel, William Lehr, Terrence McGarty, Milton Mueller, Michael Noll, Jom Omura, Robert Pepper, Peter Pitsch, Ken Robinson, Aaron

Rosston, Andrew Schwartzman, Nadine Strossen, and John Williams provided helpful comments.

1. Paul Baran, "Is the UHF Frequency Shortage a Self Made Problem?" Paper presented at the Marconi Centennial Symposium, Bologna, Italy, 1995; George Gilder, "Auctioning the Airways," *Forbes,* April 11, 1994, 99.

2. Indeed, there is much new high-frequency spectrum to open up, and much old spectrum to use more efficiently.

3. The problem is similar to that of pricing necessity discussed for the currently "free" Internet system as it is experiencing congestion problems. See Jeffrey K. Mackie-Mason and Hal R. Varian, "Economic FAQS about the Internet," *Journal of Economic Perspective* 8 (1994): 75.

4. *Sable Communications of California Inc. v. FCC,* 492 U.S. 115 (1989).

5. Concern with effects of auction or services prices was raised by the European Commission in a Green Paper. Commission of the European Union, "Towards the Personal Communication Environment," DG XIII, January 12, 1994, p. 26.

6. Howard Shelanski and Peter Huber, "The Attributes and Administrative Creation of Property Rights in Spectrum," Paper presented at the Conference on the Law and Economics of Property Rights to Radio Spectrum, Marconi Conference Center, Marshall, California, September 1996.

7. Different frequencies have different characteristics, which make them suitable or unsuitable for certain types of applications. Some of the frequency characteristics are in-building penetration of the frequency, antenna size, cost of radio components, effects of atmospheric and climatic conditions, usable coverage, bandwidth, and speed of transmission.

Part IV

Information Have-Nots and Industrial Policy

12. Why Government Is the Solution, Not the Problem: In Pursuit of a Free Market for Telecommunications Services

Gigi B. Sohn

I am often asked by free-market scholars to debate whether the government should have any role in ensuring that all Americans have affordable access to advanced telecommunications networks and services. Those scholars argue that government should have no such role, that it should stay out of the technology business, and that the market will determine who gets how much.

But that all-or-nothing approach ignores current realities and is ultimately counterproductive. Libertarians must get past the notion that government is one day going to either disintegrate or legislate itself out of existence. And they must accept that subsidies benefiting the rich and the poor will not disappear overnight. Thus, the important questions become, not whether government should be involved, but *how* it should be involved and how that involvement can help libertarians reach the goals of smaller government and freer markets.

I would argue that government can play a constructive role in making markets work better and thereby lessening the need for government involvement in the future, and, in particular, in obviating intrusive content-based regulation. It can do so by ensuring that all Americans have access to the tools that are becoming more and more central to education, the economy, social interaction, First Amendment values, and democracy. And it can do so by making more competitive markets that are currently dominated by entrenched monopolies.

Gigi Sohn is the executive director of the Media Access Project.

Affordable Access to New Technologies Makes Markets Work Better

In the age of the Internet, nondiscriminatory and affordable access to telecommunications services is needed more than ever. New technological advances—and those that are yet to come—will redefine telecommunications services and increase their importance well beyond that of ordinary telephone service. The technologies will bring new modes of exchanging opinions, information, news, and viewpoints; new tools for education and skill development; new methods for conducting research and commercial activity; and new means of engaging in civic discourse. More than ever before, the technologies can promote important First Amendment and democratic values.

Ultimately, affordable access to new telecommunications services increases efficiencies that help markets work more efficiently and equitably. New technologies improve the distribution of information that is essential to participating in the market. Affordable access equalizes the receipt of the information, thereby erasing many of the inequities that exist today when some people in the market are privy to greater intelligence than others. And when all speakers—commercial as well as those with political messages—can reach all individuals, the market will function even more efficiently.

When everyone has greater access to connected individuals, reliance on government decreases and democracy flourishes. When people use advanced telecommunications services for education, research, and development of job skills, the economy benefits and the burden on government job training and welfare programs lessens.[1] When people use the services to receive news and information, they become more empowered to make informed choices at the polls and to contribute to civic discourse and to the American system of governance. When they use the services to access public safety and health care information and assistance, they can fight crime and receive treatment for health problems.

What Kind of Access Will It Be?

Access to the Internet and other advanced networks is essential to ensure full citizen participation in society. Those networks are the backbone of the information age and will soon become the principal means by which we communicate with each other. But unlike Newt Gingrich (who later apologized for his remarks), I am not advocating that government provide "a computer in every home." Equipment is inexpensive and often free—witness the greatly publicized donations of computer hardware and software to schools and libraries by Apple and Microsoft. Nor is the latest hardware necessary to access the Internet and other advanced networks. What costs money—big money—is access to the networks, whether by telephone wire, satellite, wireless, broadcast, or other technologies. But there are no newspaper headlines about telephone companies or other telecommunications providers giving away that kind of access.

It is in ensuring free or low-cost and affordable access to advanced networks that government can play a constructive role. Such access should be provided broadly, not just to schools and libraries, which have long been the favored recipients of both corporate and government largesse. But by supporting "affordable" access, I am not advocating "equal" access, or even the best available access. Nor should the government mandate a particular technology or "pipe" through which access to advanced networks is provided. Instead, the government should ensure "adequate" access, enough to enable an individual to participate in both the economic marketplace and the marketplace of ideas.

Affordable Access to Telecommunications Services Is Needed Because a Free Market Does Not Exist

The common belief of free-market theorists is that, unburdened from government regulation, the market will determine the optimal level of technology necessary and deliver it to those who need it. As to those with limited resources, the argument goes, Kmart and McDonald's serve the poor, so why cannot technology companies do the same?

The analogy fails for two fundamental reasons. First, the telecommunications industry faces an entirely different cost structure and revenue expectations. Businesses serve customers where there is a profit to be made, that is, where revenues are greater than costs. McDonald's and Kmart may make profits from the poor and rich, but telephone companies do not often do so in many rural and poor areas of the country.[2] That is because it is enormously expensive to lay telephone wire in rural areas, a cost that is rarely offset by telephone usage. In poor urban areas, per capita costs may not be as high, but neither are revenues, particularly for advanced services. Thus, it is not difficult to find certain parts of the country where telecommunications companies have refused to deploy advanced services. For example, Regional Bell Operating Companies (RBOC) proposals to deploy "video dialtone" in the early 1990s did not include inner cities. And the cable industry's cable modem experiment in schools across the United States benefits mostly upper-middle-class suburbs.

Second, unlike the fast-food and discount retail markets, the market for local telecommunications services is not now competitive. Nor will it be made competitive just by announcing that "the local telecommunications market will now be competitive." The Telecommunications Act of 1996 has not begun to eliminate the discontinuities left after years of monopoly dominance by the RBOCs, and it is unlikely ever to do so. Because the act failed to provide a strong mechanism to require RBOCs to make their facilities available at reasonable prices, potential competitors are finding that the costs to enter the local telephone market are extremely high and are therefore proceeding very slowly, if at all. Indeed, AT&T, with the deepest pockets, believed it could compete only if it merged with an RBOC; hence its failed attempt to combine with SBC Communications. And MCI's proposed merger with British Telecommunications met with great difficulty because of the escalating estimates of the costs of competing for local service.

Even if those companies and others do eventually provide some competition to the Regional Bell Operating Companies, few believe that such competition will affect most residential users. In the meantime, according to a Bell Atlantic-backed study, there are inner cities in the United States where home telephone penetration is about 50 percent. Thus, the most fundamental entry point to advanced networks is not available to many poor Americans. That some lack access to even a basic telephone line renders irrelevant the argument that Internet access is inexpensive (although mostly in cities).[3]

What is the answer to the lack of competition in local telecommunications markets? Insistence on the old libertarian standard of "less regulation is better" simply will not lead to more competition and a freer market. Wishing the government away will simply lock in the inefficiencies of an anti-competitive monopoly. The 1996 act proved that. Instead, government can act to break down economic barriers and discontinuities in the local market by devising strict requirements that RBOCs open their networks to facilities-based competition. And if, with government assistance, competition does ever develop for telecommunications services, a phaseout of government subsidies could be appropriate.

The Benefits of Ensuring Nondiscriminatory Access Outweigh the Occasional Misplaced Government Subsidy

Even those who support government subsidies would not argue that they are perfect. They do occasionally benefit the undeserving along with the deserving. The current "universal service" scheme that subsidizes rural residents occasionally (and far more rarely) benefits the wealthy rancher or Aspen ski buff. I would prefer targeted, needs-tested subsidies, such as those the Federal Communications Commission developed for school and library access to advanced services. Under that program, the neediest schools and libraries receive the most funds. And, while we may not agree on the benefits of subsidies, I agree strongly with libertarians that subsidies must not be hidden. If government is going to give out money, it should be subject to public disclosure.

But even in the event of the occasional misplaced dollar,[4] the benefits to society that redound from affordable access to telecommunications technologies outweigh the detriments. As noted above, access to technology actually *decreases* dependence on government by providing opportunities for education, jobs, civic participation, and health care. And individuals, institutions, and businesses all benefit from a telecommunications system where each is connected to all. What good would your telephone be if there were a significant segment of the population that you could not call? The need for similar externalities for new telecommunications networks increases daily as larger and larger segments of the population rely on them to communicate.

New technologies will reduce the public's reliance on old forms of media for their news and information; that is also a net plus. Mass media consolidation continues unabated, and it is predicted that in the not-too-distant future a handful of large companies will control most of the mass media in the United States. In addition, many of those large media companies are involved in joint ventures with each other, which reduces any incentives they have to compete vigorously with one another. That concentration of ownership has reduced competition and displaces independent voices, thus decreasing diversity of viewpoints. It often leads to a type of corporate censorship by which information affecting the large media company's economic interest is kept from the public's eyes and ears. By contrast, those speaking in cyberspace can voice their opinions to many or few, with little fear of gatekeepers. That is not to say that those networks will ever replace the mass media as the primary and most pervasive source of political speech, national and local information, and news for Americans. But they will continue to have a bigger and bigger influence on public opinion and culture.

But perhaps the greatest benefit to society is what will be *avoided* if affordable access is ensured. Nearly 30 years ago the Kerner Commission described the alienation, social unrest, and other negative effects resulting from the lack of minority access to the mass media. The commission found that the dearth of positive portrayals of, and news and information about, minorities contributed to their disenfranchisement—a situation that continues to affect society today. Access to the best education, well-paying jobs, and the ability to communicate and participate in democracy increasingly depends on access to new technologies. The gap between the information rich and poor will only be exacerbated if those critical tools are available only to those who can pay a premium for them. If that chasm is permitted to grow, the libertarian dream of free markets and smaller government will be subsumed by it, as political pressure to correct the resulting inequities will overwhelm market forces.

Notes

1. A 1994 study conducted by the Children's Partnership estimated that in the year 2000, 60 percent of new jobs in the United States will require information and technological skills.

2. It is unlikely, however, that either retailer would find it cost-effective to open a store in rural Montana or in deepest Appalachia.

3. Currently, competition among Internet service providers results in low Internet access costs in some parts of the country. But government fiat enables this competition; the FCC has exempted Internet service providers from the 6-cents-per-minute access charges that RBOCs charge long-distance carriers for using their networks. If the RBOCs succeed in having the exemption lifted, the likely result is that small Internet service providers may go out of business and prices will fall more slowly.

4. Subsidies for advanced telecommunications services are similar to others that benefit the public. For example, public transportation, public roads, and highways are subsidized to the benefit of the rich and poor. I believe that subsidies that ensure access to telecommunications services are even more important because, unlike the others, they affect the extent to which members of the public can exercise their First Amendment rights and participate in our democracy.

13. How Can the "Information Have-Nots" Be Helped?

Lawrence Gasman

Politicians, regulators, journalists, and analysts of every possible political persuasion believe there exists a group of people, "information have-nots," who are being deprived of the benefits of the information society. Such commentators urge that it is high time that something be done about society's latest victims. And, of course, it is government that is supposed to do that "something." That argument found its most detailed expression in the universal service requirements of the 1996 Telecommunications Act and in the Federal Communications Commission's interpretation of the act.

In a policy analysis recently released by the Cato Institute, I analyze the wide-ranging support for universal service, finding many of the program's supporters more self-serving than altruistic.[1] I also find the supposed justifications for all aspects of universal service seriously flawed. Here, I summarize my views on the probable motives behind support for universal service and examine one of the most crucial universal service issues—affordability.

Support from Everywhere

Though the universal service mandates in the 1996 act emerged from a Republican Congress and were translated into regulation by a New Democrat's administration, those of almost any political stripe can find something to like (or dangerous to oppose) in universal service. The creation of a large pool of money in the form of a universal service fund has inevitably attracted those hungry for money, a fact that helps to explain the approximately 40,000 pages of public comments filed before the FCC's and state regulators' Joint Board on Universal Service. The smell of money has won friends

Lawrence Gasman is a senior fellow at the Cato Institute and president of Communications Industry Researchers.

for universal service among both traditional supporters of universal service on the left and their supposed enemies in corporate America. Many in the "public interest" community also support universal service for purely ideological reasons. Their somewhat utopian demands increasingly go beyond the traditional economic goals of universal service. Thus Coralee Whitcomb, president of Virtually Wired, worries that "the disadvantaged would be 'railroaded' off the information mainline" and appeals for assistance in connecting battered women to the Internet. She is particularly concerned that those in homeless shelters are unaware of the Internet's value.[2]

But important support for universal service also comes from the communications corporations that are fearful of losing the corporate welfare component of universal service. Thus, in disputing components of the FCC's Universal Service and Access Reform orders, Ronald E. Spears, vice president of Citizens Communications, testified that "we believe rural customers—ours and others—deserve the very best when it comes to communications services." Spears also stated that "the FCC's action will make it impossible for Citizens, and other rural telephone providers, to continue with the current level of investment in the network infrastructure of rural America."[3] Those statements, while doubtless sincere, coincide neatly with a desire to preserve subsidies to Spears's own company. Given Spears's position, such a concern is probably proper. But it begs the question of whether subsidies for rural infrastructure can be justified in a more philosophical sense.

Support for universal service does not come just from an unholy alliance between leftists and communications corporations. Republicans—sometimes quite conservative Republicans—are also for it. For example, the chairman of the Senate commerce committee from which the Senate's version of the 1996 legislation emerged was Republican Larry Pressler from South Dakota, a rural state, who had concerns quite similar to those of Spears.

Universal service has even been a vehicle for raising taxes by stealth. Asked why subsidies for Internet access for schools should not be provided through general taxation, then FCC chairman Reed Hundt gave an illuminating answer. Though admitting that telephone companies would pass universal service charges on to phone customers, Hundt said that that really did not matter because

> it'll be passed on to everyone in America in insignificant ways down to, you know, pennies per day. It will be a

> collective action by all America. . . . Probably the most equitable way that you could raise money for a national purpose would be through contributions by communications companies, because they cover the whole country.[4]

Hundt's statement clearly reveals that he envisions universal service as a means of levying a tax on the American people.

Expedience or Principle?

The commitment of so many different parties to universal service helps explain why the debate over universal service has been mostly about the distribution of the fund or about highly technical cost models, not about whether universal service programs are justified. But except on grounds of political expediency and commercial self-interest, can universal service really be justified?

A good start in answering the question is to take a look at who the "information have-nots" are actually supposed to be. The 1996 act identifies three groups as especially "deprived" in this regard: those with low incomes, those who live in rural areas, and certain educational establishments. The 1996 act attempts to help the groups by extending the traditional universal subsidies and creating new entitlements—notably special discounts for schools and special funding for both educational and telemedicine projects.

Supporters of universal service say that subsidies to those groups are necessary because the subsidies will bring widespread access to information technology resources, without which our society will become increasingly dysfunctional—the "general welfare" will be threatened. There is clearly some truth to all of this—but not much. Access to information technology resources will be increasingly important. But support for universal service does not follow from that fact.

A Critique of the Affordability Argument

Multiple arguments are made in defense of universal service programs, but "affordability" is at the core of all discussions of universal service. The concept of affordability, however, must be critically examined.

Currently, 93.9 percent of U.S. residences have telephone service, and 98.3 percent have televisions.[5] Approximately 75 percent of television viewers subscribe to cable television or a satellite service.

Such data suggest that, from an objective standpoint, most Americans find communications services quite affordable.

Unfortunately, however, in deliberating about the meaning of "affordability" in the 1996 act, federal regulators chose to introduce a *relative* component to the definition of affordability. Now not only must services be affordable; *all* consumers must be able to bear the cost of the services without serious detriment. That ensures that universal service becomes forever a mechanism for income redistribution; there will always be someone who must make a hard choice between telecommunications services and some other purchase.

Frustratingly, Congress failed to make its intent clear in many sections of the 1996 Telecommunications Act. But common sense helps to interpret the universal service sections of the act. A service with a penetration rate higher than 90 percent should be considered affordable without any further argument. Also, the 1996 act requires regulators defining the services to be supported by a federal support mechanism to consider whether the services

> (a) are essential to education, public health, or public safety;
> (b) have, through the operation of market choices by customers, been subscribed to by a substantial majority of residential customers;
> (c) are being deployed in public telecommunications networks by telecommunications carriers; and
> (d) are consistent with the public interest, convenience, and necessity.[6]

Many commentators, including the U.S. Telephone Association, the National Cable Television Association, and the Georgia Public Service Commission, believe that services designated for universal service support should meet *all* those criteria.[7] Any service that met all four criteria might reasonably be assumed to be a basic requirement of life. Anyone who could not afford such a service therefore might reasonably be supposed to be in need of some support—although there could still be an argument as to whether the support should come from government or from private charity.

Unfortunately, most regulators chose to interpret the list of criteria for support under the act in a disjunctive manner, opening up a far greater range of services for support and providing grist to the mill for those who want to see universal service become a full-fledged redistributionist program. Indeed, subsection (d) standing alone could be used to justify supporting subsidies for almost any kind of service. Given that affordability is now defined as affordability without hardship, the more advanced telecommunications service becomes, the more people would have to be supported by universal service. Such a possibility is hardly consistent with the view of universal service as bringing essential service to the poorest people in society. If Congress intended that result, one wonders why it bothered to list the other three criteria along with subsection (d).

To date, regulators have not pushed the redistributionist aspects of the universal service rules to their limit. But the FCC's suggestions concerning subsidies to schools have sometimes been expansive to the point of being completely outrageous. Thus, at one time the FCC suggested that schools and libraries wishing to provide high-quality, full-motion video would require a DS3 line, a 45 megabytes per second (Mbps) private line that costs the user thousands of dollars per month to operate.

The absence of any clear limiting principle in the FCC's order leaves us with a definition of "not affordable" that includes items that are not affordable in much the same sense as a Mercedes-Benz is not affordable to most of the population.

The absence of a limiting principle in universal service may stem from the paucity of the universal service debate itself. The debate placed a great deal more emphasis on which services should receive universal service support and on the mechanisms for providing that support than on who really needs such programs. There has been virtually no discussion about why such programs should exist at all.

A serious debate about why we should have universal service would have revealed that the lack of basic communications for the poor is grossly exaggerated. Indeed, 87.1 percent of households with annual incomes less than $10,000 (12,254 households total) have telephones.[8] Most of those in the group with phones—58.3 percent—also have cable television service, and 13.1 percent also have cellular phone service.[9] Cable and cellular service are not subsidized at all, and very few low-income households rely on (or even know of) low-income subsidy programs.

Even for those in the under $5,000 per annum category, the telephone penetration rate is a little over 75 percent.[10] In both cases the penetration rates are clearly below those for the population as a whole, but for most consumer products a 75 percent penetration in the poorest of homes would normally be considered a major success, rather than a major failure. Difficulty in paying long-distance charges is the main reason that low-income residents lose their phone service, not inability to pay for local service. The next section addresses whether such penetration has only come about through regulation.

The poor do spend a higher proportion of their income on telephone service than do the rich, but basic telephone service still represents a tiny fraction of the income of even low-income consumers, who spend approximately 1 percent of their incomes on telephone service.[11] Such data suggest that, although telephone service may be an essential, it is an essential that virtually any consumer could afford by making minor changes to his or her spending patterns.

Spreading Technology: An Alternative Policy

Some people contend that telephone service is so widespread *because* of universal service, and penetration would decrease if universal service supports were removed. But that presumes that only universal service supports keep prices low.

There are other factors at work here, too. In the long run, innovation and competition, not regulation, are the best and surest way of keeping prices and costs low. Penetration of Internet access, for example, is growing far, far more rapidly than telephone penetration ever has—even though computer equipment markets are not subsidized, prices are driven down sharply by competition. Indeed, telephone penetration did not begin to grow rapidly in the United States until competition between telephone companies became possible, when the Bell Companies' first patents expired in 1893 and 1894.[12]

But competition is unlikely to take off until subsidies and price controls are abandoned. And evidence is that even low-income consumers could absorb price increases with little impact on penetration, particularly because telephone service is so low a percentage of their expenditures overall. In the 1980s the subsidy from long-distance to local services was reduced through the imposition of a "subscriber line charge" of $3.50 per month, so that from 1984 to 1990 telephone rates increased—but penetration also increased, and the number of households without phone service decreased by 1.1 million.[13]

The Network Externality

Since the decline of Keynesian economics, subsidies and price controls make even left-leaning economists uncomfortable, so it would probably not be hard to convince the world that information technology would spread faster and further without such subsidies and controls. But defenders of universal service fall back on a rather curious argument known as the "network effect" or "network externality" argument.[14]

The "network effect" argument holds that the value of a network to any given subscriber is proportional to the number of subscribers on that network. That is because the more people are on the network, the more people can be reached on that network by a given subscriber. According to the argument, it follows from this that subsidizing low-income subscribers to bring them onto the telephone network benefits not only low-income groups but everyone, because there will be more people for anyone connected to the network to talk to.

That is a widely accepted argument that makes some sense. I grew up in England at a time when telephone penetration was quite low and when the poorer segments of the community would not normally have a phone. Thus, although my parents had a phone, several of the people with whom they wished to communicate regularly did not. If some program had taxed my parents to provide phone service for the poor, my parents clearly would have benefited to the extent that people they wanted to speak to would then have had phones installed in their homes. But whatever the benefits of such a program may be, the network effect argument is not an example of a market externality, because there is no evidence that, broadly speaking, everyone would benefit to more or less the same degree. That is, the benefits of adding more people to the network clearly seem to be internalized by those added and their frequent contacts, not external at all.

Thus, to continue with my autobiography, my parents would have benefited from the network effect only to the degree that a few people they knew would have been available on the phone, but other existing phone users whose social circle consisted predominantly of those who had not previously bought into phone service would have benefited a great deal more.

The Market Alternative

As we have seen, the universal service program is big business and getting bigger. A more enlightened approach than turning universal service into a major social program might be to consider direct subsidies to impoverished individuals—or an expanded role for private charities.

We should begin by applying common sense to the issue, first, by recognizing that it makes little sense to strive to eradicate "inequality" that is a natural consequence of location. Thus, it is true that rural consumers usually do have less adequate telecommunications services than do urban consumers. But there are many benefits to living in the country. Given that, does it make sense for urban dwellers to subsidize the telecommunications of rural dwellers?

One possible answer is that telecommunications and information technology is now so important that everyone must have access to it to live reasonably productive lives. However, again, some common sense is needed. Internet access and distance learning could indeed be very useful to many school kids, but should that really be the focus of education reform?

Above all, not nearly enough attention has ever been paid to the idea that the market will provide for those groups who the supporters of universal service claim are currently underserved in cyberspace. Yet, for example, innovative companies are pouring billions of dollars into developing new ways of cost-effectively supplying advanced telecommunications services to rural areas. That is happening because there are even more billions of dollars to be made by serving those customers. Some of what is being done is truly entrepreneurial and involves the highest of high technology. For example, a company called E/O Networks has proposed a plan that it calls "fiber-to-the-farm." It is a fiber distribution system intended for rural telephone companies. Since 1995 that product has been in use by rural telephone companies in the Midwest and in Canada. It is capable of delivering video and data services as well as enhanced voice services over distances of up to 75 miles. E/O claims that the cost of the system is no greater than that of an equivalent copper system.[15]

Finding new ways of serving lower income people is also an area of investment. Consider New York City, where recent immigrants often lack telephone service, perhaps because they lack a credit history or have had their service cut off for making overseas calls for which they are unable to pay. Since early 1996 a company called Microtel Communications has been providing telephones to such customers and other New Yorkers who do not get them from Nynex. Microtel allows its subscribers to make local calls and to reach directory assistance and 911 service for $23 per month. When customers want to make calls outside the metropolitan area, however, they must pay in advance at one of the Microtel centers established throughout the city. Having paid for their calls, they can talk until their money runs out. With that unsophisticated system, Microtel has used market forces to satisfy a need that advocates of universal service would insist can be solved only through government interference.[16]

And distance learning and other educational initiatives by private industry are also experiencing a boom. There is little doubt that private communications companies have made a strong commitment to providing technology to the educational sector, if only for public relations purposes. Continental Cablesystems (now part of US West) has made a commitment to supply free cable modems to schools in areas that it serves. Microsoft and America Online, respectively the world's largest software house and Internet service provider, have both offered free Internet equipment to schools. Microsoft says that its offer has been taken up by more than 5,000 institutions. Then there is Apple, which predicated much of its early marketing strategy on giving computers to the schools. The strategy of the companies mentioned indicates that private-sector firms are ready to install educational technology out of self-interest and do not have to be forced or bribed to do so.

Coda

The real reason government programs for helping the information have-nots are pervasive and enjoy so much support is not because they make any economic or moral sense but because they are politically profitable. Thus, leftist Democrats can be seen to be doing something for the poor, while conservative Republicans can be seen to be doing something for their rural constituents.

In the short term, there is little likelihood of change in the rules governing universal service. Even a purely free-market FCC would be limited by Congress, and (sadly) Congress in the 1996 Telecommunications Act clearly intended universal service to be broad in its scope.

Change must come from Congress. And growing discontent with the practical effects of the 1996 Telecommunications Act may eventually lead to its overhaul. Technological developments are already rendering outdated some of the assumptions of the act, a fact that may help build a consensus for revisions.

What remains troubling is how little the FCC, state regulators, and Congress understand or trust market mechanisms to reduce the need for universal service subsidies. There seems little reason to think that in a competitive market companies could not or would not seek higher profits in high-cost areas by deploying technology that would reduce the costs of serving those areas.

Only critical analysis of the existing law and the philosophical assumptions behind it will pave the way to a new and better—or dismantled—universal service policy.

Notes

1. Lawrence Gasman, "Universal Service—The New Telecommunications Entitlements and Taxes," Cato Institute Policy Analysis no. 310, June 25, 1998.
2. Quoted in Capital Research Center, "Cyber Activists and the Communications Revolution: Looking for Handouts on the I-Way," *Organization Trends,* July 1996, p. 3.
3. Quoted in "On & About AT&T," *Edge,* June 9, 1997.
4. "Notable & Quotable," *Wall Street Journal,* June 23, 1997, p. A14.
5. Telecommunications Industries Analysis Project, "Calculations and Sources for Revving up the Communications Economic Engine: Household Services, Monthly Bills, and Barriers to Competition," July 20, 1997, pp. 3, 7 (figures for telephone penetration are from November 1996; figures for television penetration are from 1994).
6. 47 U.S.C. Section 254(c)(1).
7. Report of the Joint Board, footnote 120. The Federal-State Board on Universal Service, Recommended Decision, 12 F.C.C. Rec. 87, 148 n. 120. (1997).
8. Telecommunications Industries Analysis Project, p. 13 (data from March 1996).
9. Ibid., pp. 17, 20.
10. Ibid., p. 14. See also FCC, Industry Analysis Division, "Monitoring Report," May 1995, CC Docket no. 87–339, 1996.
11. See Joint Board Report, page 208.
12. John Thorne, Peter W. Huber, and Michael K. Kellogg, *Federal Broadband Law* (Boston, Mass.: Little Brown, 1995), p. 797.
13. Peter Pitsch, *The Innovation Age: A New Perspective on the Telecom Revolution* (Indianapolis: Hudson Institute, 1996), pp. 77–78.
14. See, e.g., *The Economic Report of the President* (Washington: Government Printing Office, 1996), p. 176.

15. See "Fiber to the Farm: E/O Network Begins Shipping Its Fiber to the Farm Product," *Edge*, November 13, 1995.

16. "A Phone Plan Is Attracting Immigrants in New York," *New York Times*, March 18, 1996, p. B1.

14. Technology versus Egalitarianism: The Universal Service Challenge

Bill Frezza

As long as humans have been able to carry clubs, they have been concerned about equality. That concern has taken many forms. From the 16th to the 19th century, concern with equality complemented the spirited defense of individual freedom as ordinary people fought against the privileges of kings, aristocrats, and slaveowners. Today, however, equality entails not freedom but the bureaucratic institutionalization of envy, as taxes and regulations proliferate to redistribute wealth from earners to others. Defenders of freedom and equal rights constantly on the defensive against charges of "selfishness" seem to be losing ground, while the bureaucracies of envy, aided and abetted by religion and modern philosophy, grow unchecked, usually disguising their true nature under the banner of "egalitarianism."

Democratic manifestations of egalitarianism increasingly take the form of political pressure groups whose leaders champion the elevation of one "unequal" constituency or another. They then skillfully leverage their position as professional advocates to manipulate the sovereign state to effect a transfer of wealth to both themselves and, sometimes but not always, their adopted charges. An excellent example is the fiasco of school desegregation in Kansas City, where a federal judge ordered the school district to come up with a cost-is-no-object educational plan, and ordered local taxpayers to fund it. Expenditures of nearly $11,700 per pupil (more than in any other district in the country) bought higher teacher salaries, 15 new schools, swimming pools, a zoo, television and animation studios, and a student/teacher ratio of 12 or 13 to 1. The gains for the black students the plan was supposed to help were—none. Test scores did not rise. Schools were more segregated than before.[1]

Bill Frezza is a general partner at Adams Capital Management.

What does that have to do with universal service? Universal service is another example of the bureaucracy of envy in action; here is a $20 billion slush fund that not only survived the so-called reforms of the Telecommunications Act of 1996 but dramatically expanded in scope. This middle-class entitlement and corporate welfare program achieved a new lease on life by attracting a fresh crop of constituents to the dole, namely librarians, teachers, and rural health care providers—the poster children of the information age. Thus a new generation of pressure groups has been empowered, gearing up to clamor for its share of the spoils.

Some critics have pointed out that the fund could be much smaller and distort the development of infrastructure less if it were funded from general tax revenues and targeted to the really needy. Others insist on the "value to society" of connecting every school, library, and outhouse to the Internet. I will not debate those issues here; as an entrepreneur in the information industry, I am not trained in the calculus of maximizing "social benefit" by redistributing other people's money.

The question I raise concerning universal service is more fundamental. The elaborate scheme of hidden taxes and invisible subsidies is inherently unstable. Our legislators and regulators have crafted a program that is destined to spin out of control. Technology has planted a ticking time bomb in universal service that ultimately must force regulators as well as the recipients of their largesse to face the contradictions they have embraced.

The Internet as a Free-Enterprise Zone

The fatal flaw at the heart of universal service is a subtle one. It stems from the almost superstitious dread technocrats have of being blamed for choking off the growth of the immensely popular new panacea we call the Internet. That is in marked contrast to the relentless fervor with which traditional common carrier telecommunications services have been stifled and regulated. Sen. John McCain (R-Ariz.), chair of the Senate Committee on Commerce, Science, and Transportation, recently wrote in a letter to Federal Communications Commission chairman William Kennard, "Extending common carrier regulation to information services such as e-mail, voicemail, and Internet access . . . would be disastrous to the growth and development of services that have flourished over the last two decades in no small measure because they were not freighted with tariffing, resale, and other obligations imposed on common carriers."[2]

Politicians do not understand the business model driving the Internet. Many rightly suspect that it sprouted from a strange anomaly—a loophole in long-established tariff structures. The Internet is not a new network, in the sense that a crop of feisty entrepreneurs went out, dug up the streets, and laid a million miles of new wires. It is more properly viewed as a repackaging and repricing of existing telecommunications carrier services, glued together with a novel switching and protocol architecture. While the architecture offers tremendous versatility and unquestioned efficiencies, the primary virtue of the commercial Internet is that it grew silently and organically, oozing up through the regulatory cracks while no one was looking.

Chief among the advantages enjoyed by Internet service providers (ISPs) is the so-called access charge exclusion. Because they were small and flew in under the radar at a critical time when the income redistribution between local and long-distance carriers was being rebalanced, ISPs were spared from paying a per minute penalty to subsidize incumbent local telephone monopolists. While that exclusion became a matter of some debate as the Internet grew, it survived because politicians were rightly afraid they might cripple the emerging medium if they abruptly imposed regulatory burdens designed to siphon megabucks out of the pockets of multi-billion-dollar long-distance companies. That would be an embarrassing problem for both New Democrats and New Republicans, given the eagerness with which they embrace the Internet as a means for reinventing government, solving our education crisis, containing health care costs, maintaining international competitiveness, empowering the disenfranchised, saving the environment, curing racism, and anything else they can persuade voters to believe.

The commercial Internet, then, has led a charmed life, developing in a virtual "free enterprise zone," unfettered by the regulatory distortions inflicted on the rest of the telecommunications market. The regulatory community finally acknowledged that during the crafting of the latest "reforms." In the words of the FCC's then chairman, Reed Hundt, "[W]e mark the beginning of a new policy for a national data network that is based on the fundamental precept that Internet services could be in a 'subsidy-free zone'—such that

Internet communication neither relies on nor gives a subsidy."[3] What a novel concept!

Arbitrage versus Artificial Pricing

The Internet does not exist in isolation. It uses the same basic transport facilities and is potentially capable of offering the same services—including plain old telephony—as the heavily regulated, taxed, and subsidized telephone network. For the universal service regime, this is not a recipe for stability. Arbitrage opportunities whereby Internet service providers can cherry pick traditional telecommunications customers, whose costs have been artificially inflated by universal service and access charge levies, will abound. That has already started in the international long-distance market, where Internet telephony and fax are being used to bypass the inflated tariffs imposed by national telecommunications monopolies. As the technology evolves and becomes even more pervasive and efficient, nothing will protect the regulated portion of the U.S. telephone network from a similar fate. As long as the Internet remains free while traditional telephony groans under universal service regulations, investment capital will pour into the former, while the business of the latter dwindles away to nothing.

Indeed, the concepts upon which universal service depends—common carriage, subscriber lines, and basic phone service—are already eroding. In fact, in the digital age, what is a phone call? With the advent of digital subscriber line (DSL) technology that can support multiplexed traffic connections, what is a subscriber line? In a mixed competitive environment, including everything from low-earth orbital switched bandwidth-on-demand satellite services to wideband millimeter-wave hybrid public and private terrestrial systems, what is a common carrier? These definitions will not be immutable. As technology evolves and the market brings forth new business models, regulators will be forced into endless fiddling to keep the complex money-laundering machine for universal service in balance. And the fiddling ultimately will not be consistent with pledges to fence off the Internet and keep it out of the regulatory morass. Regulators will have to choose between a thriving Internet and the continuation of the hidden tax machinery of universal service.

The stock response of defenders of our current political process is, we will muddle over that bridge when we come to it. Senator McCain notes that "the advent of commercially available 'Internet telephony' services suggest a possible convergence, in the future, between information services and telecommunications. It would be grossly premature, however, to attempt to address concerns about such services today, given their early stage of development."[4] At least, the endless opportunities to campaign for reforms to the reforms will not leave the politicians, lobbyists, and bureaucrats with nothing to do—the one unthinkable outcome of any regulatory process.

So welcome to the future—the beginning of decades of old-fashioned mud wrestling, part of the never-ending theater we call public policy. And imagine an armed and surly mob of rural home-owners, underpaid librarians, and teachers' union shop stewards prepared to knock you over the head and yank a buck out of your wallet every time they want to make a phone call.

Notes

1. See Paul Ciotti, "Money and School Performance: Lessons from the Kansas City Desegregation Experiment," Cato Policy Analysis no. 298, March 16, 1998.
2. John McCain, Chairman, Senate Committee on Commerce, Science, and Transportation, Letter to William Kennard, March 16, 1998, p. 4.
3. Reed E. Hundt, "Statement of Chairman on Universal Service and Access Reform," May 7, 1997.
4. McCain, Letter to William Kennard.

Index

About the Editors

Solveig Singleton, director of information studies at the Cato Institute, specializes in privacy policy, encryption, and telecommunications law. She currently serves as vice chairman of publications for the Telecommunications and Electronic Media Practice Group of the Federalist Society for Law & Public Policy Studies. Her articles have appeared in the *Journal of Commerce*, the *Washington Post*, the *Philadelphia Inquirer*, the *Washington Times*, the *Wall Street Journal*, *Internet Underground*, and *HotWired*. Her undergraduate degree is from Reed College, where she majored in philosophy. She graduated cum laude from Cornell Law School and worked for two years at Kellog, Huber, Hansen, Todd & Evans. She lives in Washington, D.C.

Tom W. Bell is an assistant professor of law at Chapman University and an adjunct scholar at the Cato Institute. Formerly director of telecommunications and technology studies at Cato, he is an expert on such topics as telecommunications deregulation, Internet law, intellectual property, and public policy for the high-tech sector. He taught intellectual property and Internet law courses at the University of Dayton School of Law prior to joining the Cato Institute in 1997. Before entering academia, Bell practiced law at Wilson, Sonsini, Goodrich, and Rosati in Silicon Valley and at Harkins Cunningham in Washington, D.C. He earned a J.D. from the University of Chicago Law School in 1993. He lives in Washington, D.C.

Cato Institute

Founded in 1977, the Cato Institute is a public policy research foundation dedicated to broadening the parameters of policy debate to allow consideration of more options that are consistent with the traditional American principles of limited government, individual liberty, and peace. To that end, the Institute strives to achieve greater involvement of the intelligent, concerned lay public in questions of policy and the proper role of government.

The Institute is named for *Cato's Letters*, libertarian pamphlets that were widely read in the American Colonies in the early 18th century and played a major role in laying the philosophical foundation for the American Revolution.

Despite the achievement of the nation's Founders, today virtually no aspect of life is free from government encroachment. A pervasive intolerance for individual rights is shown by government's arbitrary intrusions into private economic transactions and its disregard for civil liberties.

To counter that trend, the Cato Institute undertakes an extensive publications program that addresses the complete spectrum of policy issues. Books, monographs, and shorter studies are commissioned to examine the federal budget, Social Security, regulation, military spending, international trade, and myriad other issues. Major policy conferences are held throughout the year, from which papers are published thrice yearly in the *Cato Journal*. The Institute also publishes the quarterly magazine *Regulation*.

In order to maintain its independence, the Cato Institute accepts no government funding. Contributions are received from foundations, corporations, and individuals, and other revenue is generated from the sale of publications. The Institute is a nonprofit, tax-exempt, educational foundation under Section 501(c)3 of the Internal Revenue Code.

CATO INSTITUTE
1000 Massachusetts Ave., N.W.
Washington, D.C. 20001